William White

Medical Electricity

William White

Medical Electricity

ISBN/EAN: 9783337366063

Printed in Europe, USA, Canada, Australia, Japan

Cover: Foto ©berggeist007 / pixelio.de

More available books at **www.hansebooks.com**

MEDICAL ELECTRICITY.

A

MANUAL FOR STUDENTS.

SHOWING ITS MOST

SCIENTIFIC AND RATIONAL APPLICATION

TO ALL FORMS OF

ACUTE AND CHRONIC DISEASE,

BY THE DIFFERENT COMBINATIONS OF

ELECTRICITY, GALVANISM, ELECTRO-MAGNETISM,
MAGNETO-ELECTRICITY,

AND

HUMAN MAGNETISM.

BY WILLIAM WHITE, M.D.,
PROFESSOR IN THE NEW YORK MEDICAL COLLEGE FOR WOMEN.

NEW YORK:
S. R. WELLS & COMPANY, PUBLISHERS,
737 BROADWAY.
1873.

Dedication.

S. J. CARPENTER, M.D.

DEAR FRIEND,—Fully appreciating the ability and zeal that you have brought into the field of medical reform, and the aid you have rendered in furnishing some of the thoughts for this work, as well as in its preparation for the press, permit me most respectfully to dedicate it to you. The self-evident truths herein elucidated will speedily commend themselves to all progressive minds ; and my hope is, that you may live to see them occupying their true position in our medical institutions the world over. For truth is *one*, and must ultimately prevail.

WILLIAM WHITE, M.D.

TABLE OF CONTENTS.

	PAGE
PREFACE	7

CHAPTER I.
Primal Condition of Matter, and Progress of Creation up to Man.. 9

CHAPTER II.
Man's Continued Progress as a Physical and Spiritual Being, Physiologically, Chemically, and Electrically considered... 15

CHAPTER III.
Harmonious Growth of the Physical, Mental, and Spiritual, necessary for the Health of Man....................... 23

CHAPTER IV.
Mind — its Influence over Man in all his Relations........ 27

CHAPTER V.
Conditions of Health and Disease........................ 33

CHAPTER VI.
Curative Agents.................................... 41

CHAPTER VII.
Philosophy of Treating Disease.................... 47

CHAPTER VIII.
Electrical Apparatus.............................. 60

CHAPTER IX.
Electrical Diagnosis and Medication............... 71

CHAPTER X.
Treatment... 78

ABBREVIATIONS AND MEANING OF TERMS.

P. C. denotes the primary current; S. C. secondary current. Primary current, galvanic current, and chemical current, all relate to the galvanic current of the Battery, or the primary of the Electro-Magnetic apparatus, and are all of the same nature; while the secondary current, induced current, and Faradaic current, all relate to that which is obtained from the secondary coil or fine wire helix.

PREFACE.

IN response to the urgent solicitations of many of our students and professional friends, and for the benefit of those who may hereafter enter upon this branch of study, the present work has been written; believing that it will supply a want long felt, and answer both as a guide to successful practice, and a source of instruction to all who may give it a careful perusal.

This work is not designed to supersede the necessity of the perusal of other works on this comparatively new and important branch of the healing art; for much has already been written on this most deeply interesting department of medical science, by men of the greatest eminence, both at home and abroad.

Having had many years' experience in regular practice, and a good opportunity to test the effects of drug medication alone, and being many times disappointed in the effects of medical agents, we know how to appreciate the value of an agent, which has, in hundreds of instances, enabled us to cure diseases when all our other medical resources had failed, during the past fifteen years of our medical use of this wonderful

agent. Yet we do not wish to be understood as discarding all kinds of drug medication, but would use every means within the bounds of human knowledge, whose known virtues would relieve suffering and cure disease,—human magnetism included. For we know it to be a law of electrical action, that the two electrical polarities are constantly seeking an equilibrium, and that the unbalanced electrical polarities in the human organism constitute disease, while a perfectly balanced or normal condition constitutes health. And one would suppose that this fact alone would be sufficient to convince the medical world, that whatever enables us to restore this harmony is an aid to nature, and a sure means for the restoration of health, and a blessing to humanity.

In presenting this work to our readers, it is not our design to follow the exact footsteps of famed predecessors, or to walk in the hard-beaten paths of former practitioners, but to grasp, so far as the human mind can comprehend the broad platform of truth, keeping for that purpose a mind open to truth wherever found, trusting that no one will try to believe the teachings of the finite mind *absolute and final*. We shall, then, try only to give our conception of the great and grand principles which govern human life in its operations, drawing freely for proof from all the sources to which we have, or can have, access.

MEDICAL ELECTRICITY.

A

MANUAL FOR STUDENT'S

CHAPTER I.

PRIMAL CONDITION OF MATTER, AND PROGRESS OF CREATION UP TO MAN.

LIFE, as a study, taking it from the lowest forms of organized matter to the highest known form, is one of the grandest themes in which man can be engaged.

We watch for its first development, and find it here shown in the gaseous substances of which the earth was first formed, a certain grand law governing and arranging the gases even then.

These little forms of matter (for even then they were composed of material) seemed to have been moved upon and worked over by an interior life, and to have had an acting, governing principle pervading each particle, a life and principle which we trace through all forms up to man.

Geologists tell us that they have now little doubt that this earth was first composed of nebulous matter. Astronomers agree with them in this opinion. They recognize also a law arranging and forming by degrees this nebulous mass into a globular form; then a chemical action taking place, and, by electrical laws, bringing together like elements, until the whole mass became one ball of fire. Indefinite periods seem to have passed, and still this burning condition existed; then came a time of change,—a time when these laws acting must necessarily produce change; and over and around this liquid ball of fire a crust was formed.

How slow the work! Nature seems never to have been in haste, but to have claimed all time for this *grand* work of life.

She seems to have kept this one great thought ever uppermost, and to have felt that she could afford to await its development.

We can find no form of life fitted for this age except the *interior life*, which still labored and produced one change after another, until, as age succeeded age, oceans were born, and then gross forms "to inhabit their slimy beds and crawl along their muddy shores." Through all this work of outgrowing life, we trace the same laws that first started vital existence. No labor seemed too ponderous for the controlling intelligence in accomplishing the one great object even *then* fore-

shadowed. One form of existence after another passed through its stage of usefulness, and then gave place to a still higher; and yet the work went on. Each period of time exhibited some new species of life,—some more refined outgrowth of the original chemical laws and primates, and at last presenting mankind. Man's life at first presents a curious study, so far as we can gain a knowledge of it. Little more than the common instincts which characterize animals seem to have been his. Yet he was more than they; for he combined with all they had, some peculiar forms of being which they possessed not. Nature's experiments never failed: she had commenced his life in the lowest life; and she had carried that particular part of her work through all, perfecting it more and more as she developed higher and higher beings; adding each time some new feature that should show her progress. In some of the lowest, which were fish, she had commenced the spinal column, that by and by was to bear erect man's head, with its lofty intellect, and power of reason. Again, in another, she shaped the internal ear, which now adds so much to his pleasure. How diligently she labored; for, as one part after another was formed, she gathered them together in some new expression, with ever something new added. Not only do we trace the work in the animal kingdom, but away back through the vegetable and mineral.

She had commenced her work by chemical and electrical laws, putting together certain chemical properties to produce certain effects, and governing them by the electrical laws of polarity. Never once did she mistake her method or its result; but each trial brought something beyond the preceding. Man's development was slow, and yet how certain. The first we know of him may be comparatively little, yet that little is enough to show us how much superior the human was to all Nature's previous works.

Instead of going quietly about, grazing, as did the animals around him, he seemed even at that early day to have felt the prophecy of the future of his race; and he sought rude pleasures, and what to him was employment. As before, each new period had added something new in the form of life; so now each generation gave a higher outgrowth to man. His pleasures took a more refined form, as did his whole being,—hardly perceptible at first, 'tis true; but, nevertheless, real. Like the animal, his instincts seemed to govern him; but, as generations succeeded, reason dawned upon him, and he began to see the use of objects around him. So limited seemed to be this dawn of reason, that his view of the use of all things was very narrow; but then, as now, he appropriated them for the highest uses in his power. How much was he like his good mother Earth! No rest for him or her, but ever an out-

growing life with both; so, as time passed, he sought higher forms through which to express this life, and a larger field than before for investigating uses. Like his mother, he tried various experiments; but, unlike the never-failing accomplishment of her purposes, as controlled by infinite intelligence, he failed often before his finite mind could grasp the knowledge sought. We see that it was only step by step, taken often unawares, that he gained power over the elements below him, and could feel himself lifted still higher. Faintly glimmering through his mind came thoughts of many things; and he labored to outwork them: he wished to know why certain plants grew, if not for his benefit; and why certain organs of his body were given, if not for his pleasure; so he ate and drank, and tried still further his experiments, until he found diseases come upon him. How he struggled on! but this was only the effect of his outgrowing life, the same as the pent forces in the bosom of his mother Earth; and he had only gained another step, as she gains her progress by belching her pent-up forces from the volcano's top. Why should he stop discouraged, so long as her example bade him hope for great results? It was not his nature: still, like her, he must try another and still another thought. No end, no rest, but in change; and that brought true rest. Perhaps we have no right to say that even the diseases he wrought out were not steps

of progress. All history, from the primeval records impressed upon the rocks of his early time, shows that he was ignorant of many things; and this unrest brought him in time much experimental knowledge. He began instinctively to seek some cure for the ills he felt; and the childlike feeling of awe and wonder (so natural to the ignorant) was the prevailing feeling with which he made use of the healing agents he sought. Incantations and charms, recited over the sick, expressed this feeling. Man's life to-day exhibits the same feeling in a less degree, simply through his ignorance of the laws of life. Yet how much more in comparison does he know to-day than in the past? He grasps at some natural law of use, and is wonder-struck at the result gained; but he is scarcely yet prepared with knowledge and wisdom enough to let this law provide him with a key to greater treasures of light. His out-growing life is only just beginning to realize the power it possesses: his growth is step by step as of old, one degree after another passed, and a finer being the result. Let the lesson come home to each soul, and so our reason remain unprejudiced to read still of the future.

CHAPTER II.

MAN'S CONTINUED PROGRESS AS A PHYSICAL AND SPIRITUAL BEING, PHYSIOLOGICALLY, CHEMICALLY, AND ELECTRICALLY CONSIDERED.

WE have taken a somewhat general view of man's first condition as an active creature of the early ages, and traced, as man, his relation to Nature, and all the grand laws that govern her. Now it becomes a duty to trace man as a physiological being, and still further, and more minutely, his likeness to all around him, proving, so far as we can, the ground taken.

We have seen the earth first a nebulous mass, arranged into spherical form by grand laws of electrical attraction and repulsion. So, after the copulation of the parents, we find the almost nebulous spermatozoa by the same fixed laws in the ovum, forming at first one cell, then many, until they begin to develop the new being. Some physiologists tell us of the rapid motion of these cells as they come together; and we mark there a likeness to our mother Earth: she, too, had motion, rapidly and constantly turning; and here she showed to her children one grand law of all life--

the peculiar effect the mother has over her child before its birth. Her great unrest, her struggles for still higher outgrowths, come to her children as a part of their heritage. We find that the Earth soon began to develope her different structures from the chemical primates she possessed, as the gases—oxygen, hydrogen, carbon, and nitrogen; and the materials combined with these, forming silicon, aluminum, calcium, magnesium, iron, potassium, sodium, and sulphur. Combining these gases and metals together, she makes numberless specimens of beauty and use. So, again, we find this spherical mass of cells in the embryotic human being commencing, as did its mother Earth, to collect from all around it in the womb (the boundary of its universe) these elements for its growth: sulphur for the albumen, fibrine, caseine, gelatine, &c.; phosphorus, with albumen and fibrine for the brain: silicon for the urine, bones, and hair; sodium, potassium, ammonium, in combination with chlorine, for all the fluids as well as solids of the body; calcium for the earthy salts; magnesium for the muscles; and iron for the blood.

The interior life of nature is ever working outward through all these elements; so in the embryo we see first what in the future is to do the work of man, as the first visible formation, viz., the brain and spinal marrow. After the soul-principle has collected the elements with which to

form these (and we can see this in all life), we find that it still goes on, like Earth, to form now the grosser parts of its temple. Dr. Hollick calls the principle which it first uses to work with "a species of attraction;" but he, after a time, loses sight of this, and is lost in wonder as he tries to find some philosophical reason for its continued development. To us it seems that *wonder* may very easily be turned to a delightful knowledge. If we still continue to trace nature here, we are told (and we have every reason to believe it true, for the thought brings its own conviction of truth) that Nature uses only the same great laws through all her physical growths, "that she knows no great, no small." Why should we not feel this a truth? We already see the same chemical elements combined with *his* form as hers. Chemical and electrical laws govern her growth: they must govern the growth of this, her child. Then, as they control and collect matter from the boundless realm of elements for her, so likewise do they for him; and, as they control all her actions immediately, so do they his. This, then, accounts for his onward progress in this stage of life. Chemistry controls the combining of the elements transmitted through the mother's tissues to the uterus; and the soul-germ through these refined electrical laws places them where they belong, and so builds its own machine, or, rather, earthly home.

After building the framework, like a perfect mechanic as it is, it proceeds to develop other parts; and here again we note, in almost every instance, the perfect imitation of the life-course of mother Earth. It repeats the same experiments that she has tried, and adds one new feature after another; the whole occupying some nine months of time, instead of ages upon ages.

After she has so refined the elements that the soul-life can express itself in human form, her labors seem easier, but still grand and beautiful!

How foolish the thought to estimate all below man as Nature's great work, manifesting the controlling influence of her laws, and then admitting him to be her last grand achievement, the epitome of all the rest, to profess to see reasons why from the first he should be exempt from these laws. The thought breathes neither of reason nor philosophy, but of confusion — chaos of mind.

True it is, that man is above all, because he possesses mind, the avenue of his soul-life, as a crown above all other life existences: yet that alone can come under different laws; he is still acted upon by the same physical laws. At the ninth month we find him physiologically fitted for a more active life; and even then we see a beautiful work. His life goes on by slow progression to manhood; and he stands before us a physical framework of some two hundred and eleven bones, covered by some five hundred muscles, containing

many organs of wonderful use and beauty, all the dwelling-place of this interior life. Religionists have told us for centuries of this life, and that it controls all the physical parts of man, but how or why they have failed to inform us; and physiologists seem to feel it beyond their province to mention the fact that there is a life within. They only go to work to try and explain the physical, forgetting that in every act they must make use of this life, and that to separate the one from the other materially, brings death or dissolution; and that the same law holds good as they deal with it mentally. They forget that the great facts with which they deal are but the real outgrowths of this interior life; and it leaves them sadly behind the times, for to them many physiological phenomena are unaccountable, mysterious, and wonderful. We are very glad to know that under this framework, or within it, are the brain, the eye with its nerves, the ear, the mouth, the lungs, the heart, the stomach, the liver, spleen, pancreas, intestines, and the organs which distinguish the male from the female; all this they tell us, explaining their different functions; but we want to know more. These stand as bare facts, unclothed and uninhabited by the interior life which makes MAN what he is.

Much of this knowledge can be gained alone from the physical expressions around him; but there is a deeper need — a need for man to know

of the controlling power within him. That this need is real, we shall try to show as we proceed. After his birth into the outer world, we find, from physiological facts, that life is still sustained from all around him, but not alone by the physical act of eating and drinking. His soul-principle is drawing in from all the beauty and grandeur of the earth about it, being fed by the soul-life of earth. Now, the question arises, whence this life, what its object, and why should we wish to know of it? If we again take a look backward at the past ages of our earth's existence, we note great order, — everything done in time, and all things seeming to grow from a life within; for chemical analysis shows us that we can resolve many things into chemical primates, and then re-organize them again, but that we can never give the life-portion or soul-principle back after we have set it free by our analysis. Again, in tracing this interior-life which as before we find in everything up to man, we see that some formations have more of it than others. This is a subject of deep importance, as we will now try to show; for upon our understanding of it depends much of life's happiness. Everything was made by God, and *man* in his own image, says the record of the past; and "into his nostrils God breathed the breath of life." Now, we know, that when this principle of life is gone from a plant of any kind, chemical decomposition takes place, and the particles are taken by some

other existence to feed it. So in the animal; showing clearly a power within, that, with our physical science, we cannot grasp.

When man and woman come together for the purpose of becoming parents, they then draw from the combined amount of this soul-life (which we call God) gathered and possessed by them, and it is transmitted to the child. It thus becomes a nucleus, to which is attracted from its surroundings (as we have endeavored to show) through the mother's life, everything necessary to build up the external form of the child. Life — all life — depends upon laws grand and beautiful. Let us not forget this: the physical laws are (*must be*) identical that control all physical growth, since all physical matter is of the same character; and so with the spiritual or soul of all things. We can see them expanded, as we go from the lower to the higher, yet trace them the same.

To show the importance of this seeming digression,—upon this subject so little considered by physiologists, so little cared for by most people,— we have noted in the vegetable world, that, by the action of certain laws which we call physical alone, malformations are produced, and abortive attempts at the perfect tree or flower. Again, let us take other vegetables of the same kind, and choose from among them the best, most refined, and those whose interior life is best expressed in the outer growth: the result is perfection of the tree or

plant. In the animal, the same law holds good, as all admit. Now, let us carry this law to the human, and apply it by the aid of human reason, and what is the result?—man harmonious, having less of disease, more of manhood, and a larger development of the God-principle within. Think of it ye whose province it is to heal the people and become teachers. Recognizing this principle in all things, the parents will partake largely of it in their own natures, and will take larger views of life's import, and give these thoughts to their children. They will implant in the physical germ a larger breath than others, who connect the thought of human propagation with physical pleasure alone.

We have noted the effects of unfavorable conditions upon the physical growth of plants, trees, and animals, in preventing the perfect expression of the interior life. Now carry this thought upward to the human kingdom. The lower orders seem to be stinted in physical development: they look meanly to the eye of one whose aspiration is for ideal beauty. There is an excessive preponderance of the merely animal propensities, deplorable to look upon. In many instances, there is not enough interior life to give energy to these overgrown material bodies. This lack of soul-life, manifested in so many forms, though sorrowful to consider, yet affords great lessons of usefulness to the thoughtful, and subjects for the study of those who now wholly ignore this soul-philosophy.

CHAPTER III.

HARMONIOUS GROWTH OF THE PHYSICAL AND SPIRITUAL NECESSARY FOR THE HEALTH OF MAN.

THE harmonious growth of both physical and spiritual elements seems to be necessary for the health of man. Let us look at this once more. If the interior or spiritual is the motive or life-power of all forms of existence (and this seems to be self-evident from the fact that when this is withdrawn from any form, decomposition and death are the result), then there are certain physical laws which its continuance with us render necessary. Many to-day realize that electrical laws control many of the phenomena of nature; but that this law of electricity is subject to many modifications, we who understand aught of its mechanical use cannot doubt; and if we reason naturally, unrestrainedly, we shall feel that it really is the modification of some electrical law which the soul within uses for its expressive growth without.

We plant the seed in the mold. The active life within acts upon the electrical moisture drawn, or negatively absorbed within it from the mold: soon it becomes so charged with this moisture, that its

life-principle becomes positive to the properties around it; and it starts outward and upward, repelled by similar positive conditions. At the same time it becomes polarized at the germ-centre; and then start the roots, which are positive to the mold, and repelled by the upper life. Thus harmoniously is one drawn by the earth, and the other part by the air. So we traced somewhat the modification of this law in the growth of the embryo. Now, we ask, does not this same law, expressing this same life-principle, sustain the plant? If not, what does? If blind chance alone feeds it, why does it not make some mistake? Soul-life, controlling the electrical law peculiar to its formation, attracts the right properties always at the proper time, and to the right place. Physiologists tell us of the elective powers possessed by the human system; but they scorn the idea that electrical law controls the organs within the human body: but let us see. They inform us that the food we eat, after being incorporated with the saliva of the mouth, and passing to the stomach, undergoes chemical changes; and that after passing to the duodenum, or second stomach, it is acted upon by bile from the liver, as in the first by the gastric juice. It is then taken to that portion of the intestines where the lacteals convey it to the thoracic duct, whence it is carried to the heart as impure blood. After being passed through the heart to the lungs, and acted upon by the atmos-

phere, it is changed to bright arterial blood, and carried to every part of the system. When pure, the elective qualities of the different organs enable them to select such chemical qualities as they need for their sustenance, conveying away at the same time those elements which they no longer need. Now, why admit that chemistry has aught to do with us? Simply because the fact is becoming too self-evident to be longer denied. We very well know that every chemical analysis exhibits electrical laws; how, then, can we separate the action of one law from the other? All our food becomes subject to chemical analysis as it passes from one department to another; and we must, if we stop to reason at all, know that wherever these chemical laws are acting, there electrical laws are also acting in unison; and we think, when the matter is sufficiently studied, it will be found that upon these physical laws of chemistry and electricity, governed by the soul-life, human growth depends from the first. But, says one, you would not convey the idea that our bodies are subjected to the same element that is so fearful in the thunder-storm? Why: we cannot make the same experiments with it that we do with mechanical electricity. Very true: we cannot. But do not let us forget that it is modified and refined very much ere it reaches us; and, when we come to partake of it, it must be fitted for our growth, and changed to meet our wants. Remember that

Nature, in her distribution of laws, knows no great, no small; and, again, that we have not, with all our boasted scientific knowledge, yet fathomed the depths of truth, as the constant development of new facts and phenomena abundantly prove. If, then, we find chemical action constantly taking place in all the changes which food undergoes, from a solid material condition to its formation into pure blood, we shall also find electrical action, for one ever accompanies the other. The philosophy of Dr. Dodds, concerning the circulation, is thus discovered to be superior to any other conceived by physiologists; for it comprehends more of truth, and satisfies the unprejudiced human reason better, than any before presented. If, then, our investigation of facts compels us to admit the truth of this philosophy, we can easily perceive the importance of this knowledge to us all. It enables us to make that which seemed mysterious before, clear " as the light of day;" and it expands our thoughts by greater growth, enabling us to people the earth with more of soul-life, and less of the sensual.

CHAPTER IV.

MIND, ITS INFLUENCE OVER MAN IN ALL HIS RELATIONS.

WE have considered man both as a physical being, and as possessing an interior or soul-life, which controls the physical organization by electrical law. There seems also to exist a connecting link, or medium, between the physical and spiritual. If this be in any degree wanting, the soul-life fails to manifest itself; and yet we cannot distinctly draw the line of demarcation between this soul-life and the connecting element which we elsewhere know by its peculiar manifestations. We refer to the element of *mind*, which, through the finer chemical laws, controls the brain of man. It requires wonderful, and seemingly almost miraculous, powers to go within ourselves, passing through the system intelligently, criticising and analyzing as we go; so many laws do we discover, modifying and controlling each other.

We seem to draw this element of mind from the great Fountain of all being. Its source being Infinite, we do not as yet grasp it, but only the phenomena of its workings. It is, however, by studying the

operations of laws that we are led to investigate that of which they are really the manifestations.

True methodical reasoning results in the surest conviction of truth; we will, therefore, take you back once more to the beginning, or as near it as we can reach. We can discover Mind, systematic in its action, at work from the first. How truly it attracted, through all its working laws, every atom to its place, and arranged so that every life-form filled its sphere and performed its work! To us it gave, as the richest of legacies, some of its own attributes; and we feel something of their deep significance and grandeur. When we retire within ourselves, and begin this interior analysis, we may learn of these laws of formation; and, though we cannot apply them to the making of worlds, we can to the forming of the soul-life's temple,—the highest form of physical existence. We have alluded to that great law of nature exhibited in every department of life, viz., the transmission of her attributes and qualities through countless ages to all her children; not even making, in great general principles, an exception of man. And we find, by careful study, some traces of these great principles in his nature. He imitates her almost precisely, giving through her laws his gifts to his offspring. To present what we desire in regard to this element of mind, we will commence by showing something of its native power over the laws by which all below it are governed.

From the act of copulation by the parents, the minds of both exert over the new being a power by them scarcely dreamed of,—indirectly by the father through the mother; and the mother through every tissue and nerve of her being. It is a well-known fact, that a strong desire of the mother for certain articles of diet, if ungratified, will produce the strongest love in the child for the same. Feelings of intense love or hatred for persons will produce the same marked feelings in the child. The mother's whole life during the period of gestation, whether she will or not, is devoted to the forming of her little one: every thought, every act, bearing life to it, through her quickening being, and in its after life she may trace her own during the nine months she bore it. Mothers, does this lesson teach you aught? Does it not suggest to you the thought that you should become adepts in every science, that you may scientifically labor for the physical and mental organization of your child, thus to prepare a fitting home for the soul that you have called from all soul-life to inhabit it? After birth, mind loses none of its powers over the physical man. Each thought of the father and mother will long give color to his acts: and, as his physical life seems to have been the last creation or outgrowth of external nature, so now his individual life is slowly formed, being connected with all other expressions of the same element, and still depending on all for sustenance. How intense the action

of this individual life in childhood and youth, taking all the vital force of the system for its growth! It sometimes produces fevers and diseases, many and peculiar to children (the seeds of which were sown by the action of the parents before its birth), sometimes culminating in the dissolution of soul and body; all because these laws of mind are not understood.

Few ever think of these things as bearing at all upon the welfare of the race; but still we trace the influence through all man's life, every emotion of the mind using its power over the bone and muscle of the physical. Why should they not? The intelligence of the great soul of all things has left its impress upon all; and its wisdom has shown that the knowledge we possess, and the attributes given us, are given as a power to be used. How plainly we note in the mind of our patient the symptoms which must govern us in our treatment for the day! Sometimes the dull stupidity must be stirred up into life; and again the extreme nervous condition soothed to rest and quietude; then, again, despondency be made to give place to lively joy and hope: all these conditions of mind acting for the patient's benefit or detriment, because the soul-life within cannot act harmoniously through inharmonious conditions.

The very fact that people do not understand these laws of mind, accounts for their remaining in these conditions.

That these thoughts and principles should be interesting to the masses is beyond dispute; but that they pay little attention to the cultivation of this element as an avenue for the expression of the soul—the "image of God"—is also true. If each individual was to try to develop the physical body, by giving due attention to every part of the organization, the result would soon be decrease of disease in every form. For, if the physical system be in any of its parts inharmoniously put together, the result will be a repetition in the expressions of the spirit of the malformation existing in the body. It will be one-sided, and produce consequently unhappiness.

Now, if the power of mind over these physical conditions could once be realized, and the knowledge become general that by slow degrees much of this unhappiness may be remedied by mental effort and control, how soon might true reform commence! We may note one prominent fact in connection with this element of mind. Unless some power of soul-life stirs it to action, it remains inert. This brings to view another fact, viz., that we possess power over others to draw them into the consciousness that these mental powers are lying dormant within them, and that they only need a little of the sunlight of action to start the germs of thought and soul-growth. We must remember that these laws of physical existence are only outgrowths of the soul of the physical: con-

sequently we may trace in the workings of the physical laws some analogy to the laws of mind, or the soul that governs mind; and in that analysis we shall find that the physical cannot exist without this soul. How necessary, then, that we attract to and combine within ourselves all possible elements of mental power! for we may well reason, if we, as parents, can so distinctly portray ourselves upon our children, without even a thought of what we are doing — then, as beings possessed of mind, the voluntary power which we possess over physical matter is illimitable; and, as our brother's keeper, we have in charge a talent that we cannot in justice bury from sight. We are in duty bound to make the highest possible use of it.

CHAPTER V.

CONDITIONS OF HEALTH AND DISEASE.

MUCH has been said, and volumes written, upon the subjects of health and disease; and very many writers and thinkers have at last come to the conclusion, that they are mysteries beyond our comprehension. But, in coming to this conclusion, it seems to us that the fact is overlooked, that our being is Nature's ultimate of all her primates, and that through all her physical being her laws are the same, growing only a little more refined as the elements of physical growth become refined. It may seem like a strange idea to many that rocks have soul-life; but, nevertheless, we believe it true. And why? because we see a time when this life-element becomes disturbed, and the atoms of rock-nature crumble to dust, and the soul-life is taken up by other and higher forms. But, say the material scientists of the day, prove this to us. Well, sirs, the very fact, that, in all your chemical experiments, you cannot endow your own chemical compounds with this element of soul, should be proof positive to you that you have not grasped its power, and its life you are not wholly masters of.

Because your science has not yet grasped and analyzed this soul-life, you should not ignore its existence.

How unnatural we become in reasoning thus! how obscure the labyrinthine path we tread! and how doubly mysterious the laws of life become to us! But why take us back to the rock? asks the reader: what has this to do with our own health or disease? Once more do we wish to impress upon you the fact, and so we repeat the idea again, that all physical laws are the same. This being the fact, it gives us a most philosophical basis for a true theory of health and disease, and a sure method of cure. We are not led through a mazy mass of theories to arrive only at the conclusion that there is no truth to be found. But naturally we reason; and the old giant *Might* cannot put out of the way the new truth *Right*.

Again, if the exterior conditions of the germ of tree or flower become disturbed, the plant grows into a dwarfed, crooked, or unhealthy condition: there is neither beauty nor symmetry in the exterior, and we dislike to look upon it. Even after the germ has produced the plant in beauty, a wound or sudden blow may leave a life-long scar, or sever at once the connection between its soul-life and physical expression; and we say the tree or flower has perished. Now, we may take the very same seed, and give the right conditions of light, heat, air, and richness of soil, keep away

from it the insect that saps its life-blood and the powers which can in any way harm or injure it, and the result is harmony and order. The soul still governs and forms its beauty; and, instead of dried branches and dwarfish fruit, we have it teeming with life and health in all its parts. So of the rock: if no elements of discord come to its growth, it retains its soul-life complete; and we trace this law through all forms of matter up to man. Why should man believe that the laws of physical life are beyond his ken, when they are so plainly written upon every chemical primate of his being? O foolish man! how little of truth you grasp with your mighty intellect and growing reason, because you simply try to become unnatural!

We have said that we can trace these laws up to man, and we further know that in man's life on earth they are the same. It is only necessary for us to throw away prejudice, and look at this subject as rational beings, and we shall see at once the truth of this proposition.

We know, that if unfavorable conditions, similar to those which we give the tree, rock, flower, or animal, be given the human mother, they will produce similar results in her offspring. Idiocy, impure thoughts, impure acts, bodily disease of the most loathsome character, disorders of the worst forms, are but the results of this one great law, as may be proven by visiting some of the

denizens of the streets of our great cities, whose very names make the pure shudder. The most scientific cannot deny this fact. Why is this? What does it mean? only that the soul-germ, the image of the Father, has not the conditions through which to express itself. One might almost say, and no doubt many *hope*, it does not exist at all in these conditions. But here, again, the great Father's law is manifest: some little act of these poor of his vineyard will show this soul-life, this interior, even here.

We cannot help the feeling of wounded self-pride that some may have, as the conviction of this self-evident truth comes home to them, that out of the dust of the earth they are formed. But why murmur at it? why try to discard the fact? Is not the lineage grand beyond all human power to give? Look at the majestic mountain, and the giant rock-crowned brow of yonder cliff; hear the swelling anthem of old ocean. Is it not a thought beautiful in truth, that from all these we sprang, and that the same grand anthem was our cradle-hymn ages ago? We have only calmly to seek our mother Nature, and listen to her voice, to know that all her laws are beautiful and true; only to ask of her, and she tells us, beyond the fear of doubt, that health is only the harmonious condition existing between this soul-life and the exterior or physical. In other words, health is only the manifest order of the soul over the mag-

netic and electrical laws, by which and through which it governs and sustains its relation with the chemical elements which compose the human body. Once place within the reach of these laws an element which they can grasp, and, if it be not chemically allied to those of the human body, this fact is sometimes proved to our sorrow; for the soul-life must gradually leave this disorder to seek for order elsewhere: hence we witness the phenomena of death.

How sad to feel the struggle of this interior life for exterior expression in the class before alluded to! So great is the disorder of the elements of the physical, that its language can only be of the lowest kind. Oaths and imprecations, obscene phrases and acts, can be its only mode of expression; for all the magnetic relations of mind and matter, and most of the electrical laws of the chemical primates, are disturbed. Through this very means we might arrange society into almost numberless grades, until we reach that class, few in number, where the magnetic relations are rightly sustained, and the chemical primates exist in their proper proportions. There we find harmony and order. Beautiful thoughts flow from such beings in beautiful language; no rough act bespeaks inharmony of any kind; but to their soul-life everything is in its place, governed by its laws, and its searchings after truth may and will benefit all.

In thus defining health, we very readily perceive that disease is its opposite, or that state of the chemical or electrical laws with which the soul cannot harmonize, consequently cannot control, disorder being the result.

That each chemical constituent of the human body should sustain a true and harmonious relation with its neighbor, and that a correct perception of the electrical laws of these chemical constituents is necessary, are truths that would seem to be self-evident to all the medical world, if they but think of and comprehend their own theories. But how much labor these thoughts imply! how necessary that the masses become thorough chemists,—that they know of what they are made, by what sustained, and the laws that govern and control every part of their mechanical and chemical composition! As it is, life is a dull study, and is left by them to a few, who, working comparatively alone, having none of the stimulus necessary to their soul-growth, settle down upon partially-discovered facts as whole truths, and therein make mistakes fatal to their future progress as men of science.

This thought alone should give a new impetus to every student in the great field of thought and science; and they should never allow themselves to be content with what their forefathers have discovered, but ever feel that truth, in all directions, is as broad as the universe, and as deep as the

fountains of life in the bosom of its Maker. That we can grasp it all, may seem impossible: yet it is only by constant growth that we develop our capacities for soul-expression. "But," says one, "what a curious conception of health and disease: if it all be true, what is our best course of action? Yet how can it be true? Old authors have never discovered anything to prove it; but all their researches seem to prove its falsity." Stop, readers, and recall the facts recorded by old authors. They tell you, that, from the moment of conception, the embryo human being is composed of chemical properties; that, from that moment, it grows by a species of attraction; that, after birth, its food is composed of certain chemical compounds, each chemical part finding its fellow within the human system, after having undergone certain chemical transformations during the process of digestion; that, when this process is completed, the system then selects its food from the blood by an elective agency; that, if these properties become disproportionate, the organs are disturbed in their functions, and also that the conditions of rest and labor act upon these organs in their agency of election; that the great power of mind may so act upon them as to change every chemical quality, producing a mass of disease terminating in death; and, finally, that there is a nervo-vital force which seems to run the whole machinery; and, when that ceases to act, the body is before us, and all its parts, but has no more power

to think, act, or be. Well, what is the inference to be drawn from all this? What can a reasonable mind conclude, but that this soul-life does act upon these chemical compounds, through electrical and magnetic laws, thus sustaining us; and that these laws are the elective agency, and the species of attraction, by which all agree we **were** made, born, and still have our being.

CHAPTER VI.

CURATIVE AGENTS.

WE may sum up the conditions of disease as the change of chemical qualities, and the change of polarities of the electrical laws which govern them; these changes rendering the harmonious action of body and soul impossible. The means of cure, then, must be based upon these two facts; and it leads us—where? Away into the fields of nature, from whence we sprang, up and through the world of nature to man: and we may not even except him; for in his composition do we behold, from the earliest days, that which may cure his brother-man of many diseases. We are now led to inquire what we must do, and how? Simply, we answer, change the chemical qualities, and restore the equilibrium of the electrical force, and you have cured the disease. But how to do it is a question that has puzzled for ages the wisest savans, because they have not followed out the laws discovered by them in the human organization. They found, and we find, all that man is composed of in nature below him, manifesting there the same inherent laws. If we think of this,

and also of the fact that with us the material takes a more refined form, we have a sure method. Chemistry has taught us, that if we enter the precincts of her domain, we may learn the art of refining in a degree the elements which she claims as her own; but that they can only undergo certain changes, without losing that soul-relation which they must sustain to us in order to do for us what we wish. That when we reach that degree, then the soul-life within each must do the rest.

We find everything teeming with Nature's life and love: as students of her school we seek, still keeping in view her laws, so far as we can comprehend them. The result is, that we can see a means of restoring this lost balance in her mechanical department, though we may hardly call the action mechanical, for it is only one of her products producing chemical change upon another, thereby evolving again the law of electricity. This we find to be a great agent in the cure of disease, when used according to the comprehensive law of electrical polarity. Its mode of action we give you in this work,—a great agent, because it finds a home within us, and reigns as governing principle in the material organism. In fact, we have seen this principle to be the governing one of all life; mineral, vegetable, and animal, selecting and placing the atoms that form each distinct form of life in their appropriate places. As we find the rocks and minerals to be grosser, to

express less of soul-life, so we find that they evolve a grosser electrical quality; and, to eliminate it for our use, we must use mechanical means. In using it through these means, we may learn to know its law as developed in all else; and we shall see also the use of the various remedies, keeping ever the thought with us as expressed by our philosophy, that after the mineral formations came the vegetable, and their structure was one step nearer the human. Consequently this fact should teach us a lesson that many discard, viz., never to scorn their use as remedies for disease, any more than we would discard them as articles of food. We should also remember, that after most perfectly preparing and administering these remedies as we should, the various organs must each select from them by elective affinity (spoken of by different school-authors, and which is nothing more or less than electrical attraction) that chemical property which contains the most appropriate amount of the magnetic or electrical quality for its use; each organ having to work over and transform the different ingredients into its own life, as with food. The process is still, in fact, digestion, after the stomach has performed its part. So with mechanical electricity: when used, each organ must digest it through its nervous power, before these life-elements can make use of it. All life seems to be but one great chemical laboratory; from the lowest life, the object of the refining pro-

cesses, and different analyses, is the development of man. But, though this is a generally acknowledged fact among many philosophers, they are not ready to make practical use of the knowledge they possess; and many who see the use, as a remedy, of electricity obtained from minerals by mechanical means, are not at all ready to look at the same life-principle in the other remedies used, considering all alike as of little importance. At least, that is the inference we draw from their philosophy and practice.

But, says the reader, you do not except even man, but declare that he possesses in his composition that which will cure many diseases: how are we to prove this? You certainly would not advise us to turn cannibals, and so get at these qualities? No: far from that, friends. Man is the epitome of all else; this fact alone is sufficient for the basis of this great philosophy. He possesses thousands of nerves, through which he sends from his soul-life this power of magnetic life. Here, again, let me call your attention to the fact, that, besides being the sum of all the rest below him, he is more; for to this soul-life he adds the attribute of intelligence and progression: so that, besides drawing constantly from all below him to sustain the physical, he is drinking in from all around to sustain the soul. While, therefore, the soul-power collects from the physical fount of nervous power the electrical element, it at the same time sends from

its own interior the magnetic, and, combining the two, transmits them rapidly from one point to another to perform its will, gathering from each peculiar organ its peculiar life-force.

Now to apply these facts,—for every physiologist will admit them, if not in this exact form,—if health be the manifest order of the soul over the physical in all its departments, then the physical must be able to transmit, by the law of order, the soul's great chemical compounds to those whose souls possess not this order. The law is as grand as it is natural and beautiful, the popular world to the contrary notwithstanding; for it is only the outworking of man's power in another form.

To apply this power, he has only to remember the law of mechanical electricity, and to feel that Nature treats her children all alike, and in the distribution of her laws she shows no partiality.

After the foregoing, what must be our conclusions? First, that man, as a being, had his origin from two sources,—all forms of matter as a physical basis, and the great intelligent Maker of all, as possessing a governing power over the physical; that he brings up from all forms the same properties, more refined, yet so much like them, that still he must use all forms to sustain his physical; that there are certain laws by which he is enabled to do this,—these laws being the chemical laws of electricity, and its polar attraction and repulsion; that, to do this appropriately, he must

thoroughly understand these laws, and so enable the interior life, the intelligent part of him, to control the physical harmoniously and with order, thereby maintaining the physical in perfect health. That disease is only a change in these chemical atoms and their polar relation; that, to restore these atoms to their proper and normal condition, we must make use of all the philosophy of chemistry we possess: first, because that is the most natural mode of reasoning; and, secondly, because only by being natural, can we be cured. That we find remedies in all Nature; and we have only to be true to her teachings, and we shall find that we can and do from her gain all the power to restore health.

And now, if we have given our readers a part of what we conceive to be the truth, and as such they can accept it, we are gratified. If not, then we ask them again to look at the philosophy of life, seeking for truth still from Nature, and then proceed with us to the chapter on the mechanical and other means of treating disease by electro-magnetism, electricity, &c.; learning from that the law of all remedies, as we learn the prophecy of all life from the mineral.

CHAPTER VII.

PHILOSOPHY OF TREATING DISEASE.

WE lay it down as a fundamental principle, that there are truly only two phases of diseased conditions, which, for the sake of being properly understood, we shall call *Positive*, or Hypersthenic, and *Negative*, or Anesthenic. Under the *positive*, we include all such as are attended with inflammation, congestion, soreness, acute pain, bruises, fevers, sprains, extraneous growths, expanded muscles, and swellings of all kinds.

Under the *negative*, are included paralysis, local or general debility, contracted muscles, nervous prostration, coldness of the extremities, torpid liver, and inaction in any part of the system, with atrophy, or tendency to decomposition, local or general.

Corresponding to the above classification of diseases, we have the *positive* and *negative* poles in our electro-magnetic and electrical machines and galvanic battery. We have discovered, that, wherever we place the positive electrode, or pole, it decreases the electrical action in the part; and, wherever we apply the negative electrode, it increases the

electrical action; and that just half way between the electrodes the current is neutral, whilst at the two extremes we have the greatest positive and negative action: or, in other words, one-half the distance is positive, the other negative. Hence, by understanding how to direct the current, we can increase or diminish the electrical action in any part at pleasure; and, as all diseases exist by virtue of an unbalanced condition of the electro-vital force or currents in our system, by restoring an equilibrium, we lessen, remove, or cure, the disease.

We have discovered, also, that the positive current is alkaline and hot, while the negative is acid and cold; that, where the positive enters, it produces a cooling effect, which corresponds to the acid condition, while the negative produces warmth, or the alkaline condition.

As all substances can be classified under one or the other of these heads, we see the philosophy of applying the electrical element, both for the restoration of the lost equilibrium, and also for the production of those chemical changes necessary to restore the different parts and organs to their normal or healthy conditions.

Again, we have discovered that the course of the current is always from the positive to the negative; hence we say the negative attracts the positive; and that the negative aggregates, or gathers, while the positive segregates, or diffuses. Therefore, if

we desire to reduce glandular or other enlargements, we apply the positive electrode to the part. It will be evident, from what has been said, that where we apply the positive electrode the current *goes in*, and where we apply the negative, the current *comes out:* hence we say all positive currents are *inward*, and all negative currents are *outward*.

Admitting that the nervous fluid is a modified form of the electric fluid, which undergoes in the brain and great nerve-centres that change which best fits it to become the agent of the mind, we see how it really forms the connecting link between the soul and all its organs and functions of life; how it becomes the life-force of the animal economy, and hence must permeate every part of the organism. It has, for its great highway and fountain, the brain, spinal cord, and threefold system of conductors, viz., the nerves of sensation, and nerves of voluntary and involuntary motion. This vital or life fluid is subject to the laws of electrical polarity, both in its general circulation and in every organ of the body. When in health, the positive and negative poles balance each other: but any agency which changes this relation may be the cause of disease; causing that to be abnormally positive which should be negative, and that which should be positive to be abnormally negative. We correct this abnormal polarization by properly applying the poles, thus curing the disease.

In order to have our currents produce the desired effect, the circuit must be perfect; hence the two poles must inclose the body, or the part to be treated, between them; which then acts as a conductor, and makes the circuit complete.

Three things are absolutely necessary in the successful treatment of disease, viz., a correct diagnosis, a thorough and scientific knowledge and application of the agent employed, and correct habits on the part of the patient. After making our diagnosis, we must, in treating our patient, pay especial regard to polarity. As was before stated, when the two electrodes are in contact with the body, one-half the distance is positive, the other half negative, and the centre may be said to be neutral.

In treating inflammatory, or positive conditions, we must bring all those parts under the influence of the positive part of the current; for, if we do not, we shall fail in changing the polarity and curing the disease. In treating all inflamed or painful conditions, it favors the operation to run the current with the nervous ramifications, because its effect is more soothing. Even in treating such cases *through and through*, it is well to keep the negative a little below the positive.

In treating sub-acute affections, we must act according to the necessities of the case; but, as a general rule, it is best to subdue the pain and irritation first, by treating with the positive over the

painful part, although in some chronic cases it may, strictly speaking, be a negative disease. As all negative diseases require increased action, the parts affected must be brought under the influence of the negative end of the current; and in many cases of wasting, or atrophy, and paralysis, we often allow the current to run against the nervous ramifications, as it is more warming and tonic in its effects. But, in this case, we must be governed by the class of nerves which are paralyzed. If it is the nerves of sensation, we must run the current from the spine to the extremities; but, if the nerves of voluntary motion, we must run the current from the extremities to the spine, bearing in mind that the current always runs *from* the positive to the negative, and takes the most direct course and the best conductors. The dry skin is a bad conductor: hence we moisten it with our moist sponge electrodes, water being a better conductor. But fresh water is not so good a conductor as salt water: hence the fluids of the body that are impregnated with salt are good conductors.

Some writers claim that the muscles are better conductors than the nerves: but, since no part of the muscles are devoid of nerves, it is hard to determine to a certainty concerning the living body. We have more reasons for believing that the nerves are the true conductors, both of the electrical and every other system of circulation. Every kind of action,—chemical, mental, and mechanical,—is,

philosophically speaking, electrical in its nature; and we are warranted in affirming that every atom, cell, and organ has its positive and negative polarities, its alkalies and its acids, and is in itself a miniature galvanic battery. This is equally true of every organ of the mind. It is also true of the body as a whole, for we know that in health the mucous membrane is alkaline, while the skin is acid. This accounts for the fact, that the centre of the body and all the organs are relatively positive, while the external, or skin, is negative. It also accounts for the very important fact, that all the electro-vital currents, in their normal state, flow from the centre to the periphery; thus illustrating the same universal electrical law, that the current always flows from the positive to the negative.

Another important principle we should here consider. The skin is the greatest eliminator of the body; and the currents which constantly flow to the surface throw out, through its untold millions of pores, those elements, which, if retained, would cause immediate disorder, as is evinced by taking a severe cold, the pores thus becoming closed, and fever to a greater or less extent resulting. What is now the condition of the electrical currents? The polarities are changed, or are alternating. If the change is permanent, we have constant burning heat on the surface; which is fever. The currents are set inward; or, in other words, the surface, or external, has become abnor-

mally positive, while the centre, or internal, is negative. The appetite fails, there is soreness all over the body, with languor, headache, and many more unpleasant symptoms. Or there is an intermittent condition; sometimes the surface is hot, sometimes cold. Nature is struggling hard for the ascendency, but the pores are too firmly sealed, and the power of reaction is growing less and less. In this critical state of affairs, what is to be done? Oh! send for the doctor. And what does he do? The lancet, jalap, and calomel, used to tell the whole story; but, thanks to progress in medical science, even our drug-practice is now more humane and more in accordance with good sense. It has at last been discovered that the system needs an addition to its vital force, rather than to have it still further reduced, and thus made a more easy prey to the destroyer.

But, it may be asked, how can electricity change this abnormal to a normal condition? We say, simply by changing the polarity, and adding to the vital force, thus co-operating with the *vis vitæ*, or life-force, of the organism. This we do by seating the patient on the positive electrode, and treating with the negative all over the body; thus establishing the normal flow of the currents from the centre to the periphery. This process opens the pores, and aids the system in throwing to the surface the morbid accumulation, cooling the surface, restoring equilibrium, and curing the disease.

This is the work of only a few minutes, or at most a day or two; for it seldom requires more than three *séances* to break up a fever, especially if it is taken in the early stages. In some cases (and generally when practicable), we place the patient in a warm galvanic bath, and apply the electrodes as before. Usually one treatment of this kind will be sufficient; but such patients can seldom come to the office, and so lose this advantage.

We said, a short time ago, that all the functions were carried on by electrical action, even the circulation of the blood. It is now generally admitted, that, by the law known as the correlation of forces, all the changes and motions in the mineral, vegetable, and animal kingdom, and in the universe, so far as we know of it, are one in essential character; that all have their origin in the law of attraction and repulsion, aggregation and segregation, centrifugal and centripetal, chemical and electrical, positive and negative conditions and relations of the different elements, each being convertible into the others; and, in the last analysis, we are forced to the conclusion that electricity, in the hands of Deity, is the cause of it all. Let us now see how the blood is made and circulated. First the food is masticated in the mouth, and mixed with saliva, which is alkaline: then it is passed into the stomach; and the gastric juice, which is a powerful acid, quickly combines with the alkali in the food, and resolves it into a pulpy mass. It is

thence passed to the duodenum, where it is mixed with the bile from the liver, which produces another chemical change, it being alkaline. The pancreas also contribute a fluid, changing it still further. It then passes through the small intestines, and the nutritious part is absorbed by the lacteals, and conveyed through the mesenteric glands into the thoracic duct; thence to the left subclavian vein, mixing with the venous blood; thence to the heart and lungs, where it all comes in contact with the oxygen of the air in the lungs, and is changed from a dark purple to a bright red color. It is thence conveyed to the heart; and from thence, by the great aorta and arteries, to the remotest parts of the system; and is then conveyed from the arterial capillaries to those of the veins; thence to the heart and lungs, again to be renewed and repeat its course through the system.

Now, what is the nature of this action of which we have been speaking? You may say it is both mechanical and chemical, and we may add electrical and magnetic also; and we claim, that, so far as the circulation of the blood is concerned, it is greatly electrical, as we will now proceed to show. In the first place, we will again state the fact, that two positives and two negatives equally repel each other. The whole organism is operated by means of the nervo-vital fluid under the control of the soul-life,—the blood being the source of this subtle fluid, the oxygen of the atmosphere supplying the

blood. The blood, being oxygenized and electrified, becomes electrically positive; and the lungs being already in that positive condition, there is a mutual repulsion: and, since the lungs cannot leave the chest, the positive blood must, especially as the lungs are being constantly supplied by negative venous blood, between which and the positive lungs there is a mutual attraction. The blood thus repelled from the lungs returns to the heart and from thence is attracted into the great aorta; from thence through the whole of the arteries; the most distant being most negative, attracts the blood onward to the extremities of the capillaries; and the venous capillaries being still more negative all the way to the heart and lungs, the blood is attracted through its whole course on electrical principles. Now, as nerves accompany the arteries and not the veins, we can see clearly that these are for the purpose of attracting the electrical element from the blood, and conveying it to the cerebellum or small brain, to be used by the mind and involuntary nerves to perform all the functions of the body.

Again, as the positive blood, laden with all the rich elements of nutrition, courses through the arteries, each organ and part attracts to itself what is needful for its support; at the same time, the waste material, or broken-down cells, is attracted into the current, and, according to its kind, eliminated by the different depurating

organs of the body, such as the kidneys, liver, and lungs. Further: it is a well-known fact, that everything taken into the system as medicine, if it does any good, does it by virtue of the electrical principle of its action, either by being attracted to diseased parts, to build up and renovate, neutralize morbid elements, or aid the organs in repelling morbid elements from the system.

This being the case, it is no wonder that so much is claimed for this most potent and all-pervading element as a therapeutic agent in every department of the healing art. We see, also, how it becomes a power, in healthy organisms, for the cure of many diseases, and that in a very limited time, compared with drug medication alone. Hence we say that all the good produced by hand manipulations, or human magnetism, is in accordance with this universal law of electrical action; and is all the more potent as the will of the operator and the *faith* and will of the patient act in harmony.

Since all diseased conditions originate in the disturbance of these subtle elements of our nature, a corresponding element is needed to restore the equilibrium; and as human magnetism and electricity is the nearest approximation to the mind-element, or nervo-vital fluid, we see why it is the most reliable and potent, when wisely and scientifically applied, for the cure of disease. We also see how, in certain mental and nervous conditions,

a word, a look, or a touch, from the healer, produces wonderful changes in the all-believing subject. This, too, is in perfect harmony with law; and it makes no difference whether these things are done by men or angels, or both combined: law, universal and unchanging, is the foundation of it all. Understanding this, and the adaptability of the element we use to diseased conditions, the process of healing is simple, easy, and natural, as we will try to show in the following pages.

In the preceding pages, we have endeavored to show that this powerful agent exists in everything, but acts with more potency in some elements than others. In the constant changes that are everywhere occurring, we perceive a gradually refining process, elements serving higher and higher uses, until they manifest their highest potency in man; for what can he not accomplish, under the direction of his rapidly expanding mind and interior soul-life? No wonder that he finds within himself the most efficient element for healing his diseases, just as his mother Nature has a self-adjusting power within herself,—as witness the earthquake, thunder-storm, and tornado. The "soul of things," the soul of man, and the being of God, are so closely allied and interblended, that we know not where to draw the line between them. Man is a microcosm of all else, both spirit and matter, and appropriates from all below and

around him whatever is necessary to sustain this relation of being. So far as we know, he is capable of infinite progress: how important, then, that his mind should be free and his body healthy, so that he may fully answer the divine end of his being, living in harmony, and ever enjoying and promoting happiness.

CHAPTER VIII.

ELECTRICAL APPARATUS.

Electrical Instruments for Medical Purposes.

OF these there are many, and of various constructions; but the assortment necessary for all practical purposes is comprised in the "Friction Electrical Machine," Galvanic Battery," and "Electro-Magnetic, or Faradaic," and "Magneto-Electric Machines."

BY THE ELECTRICAL MACHINE, we get a peculiar form of electricity for medical purposes, having great intensity with but small quantity. We esteem the plate machine, of twenty inches or more, the best. To operate this machine successfully requires a dry atmosphere, and the machine to be kept free from dust, and everything kept dry and clean about it. In moist weather, or when a storm is approaching, and the atmosphere is negative, a little sweet oil rubbed on the plate is an advantage. Amalgam powder of zinc, or deuto-sulphuret of tin, must be, now and then, applied to the surface of the rubbers. A good conductor with the *moist* earth is also necessary. It is well to bear in mind that the prime conduc-

tor is always positive, while the rubber is always negative; and the one is never evolved without the other. In the Leyden jar, the inside is positive, the outside negative; and, if we wish the negative charge, we must have the prime conductor communicate with the outer covering of the jar. But the positive is the one generally required for medical purposes. In discharging the jar, always bring the knob in contact with the outer side first, and then with the knob on the top of the jar. In charging the jar, the outer foil must be in communication with the conductor to the earth, so as to dissipate the negative electricity from the outside of the jar, while it is being charged inside positively. By placing the patient on an insulating stool, he can be charged with positive electricity while in contact with the prime conductor; or with negative electricity by contact with the rubber of the machine. Where we wish to employ both electricities at once, we must not make either the rubber or the prime conductor communicate with the ground, but keep all well insulated. Having thus put the machine in order, we are prepared to use it according to the instructions for treating diseases in the succeeding part of this work.

THE GALVANIC BATTERY.

This is an arrangement by which we obtain the primary current of electricity; which is capable of

producing the most extensive magnetic, chemical, and calorific effects. By this arrangement, we get a current of large quantity with small intensity. This current, as will be seen by the following application of it to diseased conditions, is a wonderful agent for good in curing many diseases that have heretofore baffled the skill of our most learned physicians.

One battery is constructed as follows: A glass jar, containing dilute sulphuric acid—one part of acid to ten of water; in this is a zinc cylinder, and within it a porous cup, containing electropion, into which is placed a bar of carbon, with a brass clamp and copper wire attached, to connect it with the zinc of the next pair; and so on to all the cups. In this battery, the carbon is the positive, and the zinc the negative; to each of which conducting wires are attached, by which the current may be applied to any part of the system. When the battery is not in use, the cups must be disconnected to prevent constant action. When the two fluids become of *one* color, and salt rapidly accumulates on the zinc, they need to be renewed; but it will generally run well for a month or more without changing. Before the zincs are used they are to be amalgamated with quicksilver, and as often afterward as the quicksilver disappears. In order to prepare the zincs for the quicksilver, they must be placed in a strong battery solution of sulphuric acid and

water, just before applying the quicksilver. In renewing the battery, it is well to place the zincs, porous cups, and carbons in clean water, so as to free them from all accumulations. The current will be weak or strong, according to the number of cups used and the freshness of the fluids in the cups.

Other batteries are constructed, which operate on the same principle as the foregoing, except that by a peculiar arrangement one glass cell is made to answer the purpose of both the glass and porous cells, and only one fluid-electropion is used. All the cells are arranged in a neat walnut case, and the connections made in such a manner that by a metallic slide on a tramway over the cells, any number of cells can be used; and besides by a simple contrivance all the cells can be elevated or lowered at pleasure, so as to take the zincs and carbon entirely out of the fluid. This battery is a late improvement, and is portable, which makes it very convenient both for office and out-door practice. The number of cells in this battery vary from twelve to sixty.

TO MAKE ELECTROPION.

Take one gallon of warm water, to which add six pounds of bichromate of potash. Dissolve as well as possible; then add four gallons of *cold* water, and keep stirring. Next add one gallon of

sulphuric acid; and, when the bichromate is dissolved, it is ready for use.

If a smaller quantity is required, make the proportions accordingly.

THE ELECTRO-MAGNETIC OR FARADAIC MACHINE.

There are a variety of machines constructed on this principle, which consists of a double helix, one of coarse, and one of fine wire, and a vibrating armature and magnet. These are the kind most extensively used for medical purposes. They have two currents, one connected with the coarse wire helix, and is called *primary*, the other connected with the fine wire helix, and is called *secondary* or *induced*. Both currents can be increased in intensity by means of a metallic rod or tube, which in some machines passes *over*, and in others *into* the helix. Some machines have in addition, also other arrangements for the same purpose. The inducing power in these machines is a small battery arrangement, which is simple in construction, and easily cleaned and kept in order.

Great improvements have been made in all electro-therapeutic appliances within a few years; and inasmuch as a descriptive pamphlet accompanies each apparatus, it is unnecessary to enter into a minute description of them here. If sulphuric acid is used, ten of water to one of the acid is the proportion. If sulphate of copper,

two ounces of the sulphate to a quart of water is the proportion.

The acid is cleaner and more convenient; but the sulphate of copper is less dangerous to clothing and carpets, etc.

MAGNETO-ELECTRIC APPARATUS.

This apparatus for medical use is arranged in a neat box, and is the only machine possessed by many medical men. This kind of electric current is produced by permanent magnets, in connection with a revolving armature. It is kept in rapid motion by turning a crank, which turns the armatures, and thus, passing the poles of the magnet, gives a rapid succession of shocks, making a current something like the current of the electromagnetic machine. Its intensity depends upon the size and number of the great magnets; upon the size of the wire wound around the armature; upon the exact nearness of the revolving armature to the tips of the poles of the great magnet; upon the number of its convolutions, and the velocity and regularity with which the wheel is turned. No fluids are used for this machine.

REMARKS ON THE DIFFERENT CURRENTS OF ELECTRICITY.

The Electrical Machine gives us that form of electricity which admits for the most part of instantaneous discharges. It can scarcely be called

a current, except as it is being received into the system by contact with the prime conductor, when on the insulated stool, while the machine is in mechanical operation. The moment that ceases, the current stops, and the surrounding atmosphere quickly restores an equilibrium between itself and the patient so charged. If we desire to charge a patient positively, we place him in contact with the prime conductor; and if negatively, in contact with the rubber of the machine. In either case, sparks can be drawn from any part of the body or clothing when so charged. This has often been resorted to, with various degrees of success, for the cure of paralysis, rheumatism, tetanus, cataract, and amaurosis, and sometimes with good effect in nervous prostration.

The Galvanic Battery gives us that form of electricity which constitutes the direct current. It is the current we most rely upon for decomposing and dispersing morbid growth, such as tumors, nodes, and calcareous deposits about the joints, stone in the bladder, and biliary calculi. It is likewise our chief agent in neutralizing morbid elements in the system by its powerful chemical action, as also all skin affections, cancerous, syphilitic, and scrofulous. To reduce swellings and inflammations, we apply the positive electrode to the part affected; and, to increase action, we apply the negative. All inflamed parts are electrically positive: and as we know it to be a

universal law in electrical action, that two positives repel each other, and induce a negative; so, by applying the positive to the inflamed part, we change the polarity, and restore an equilibrium, or aid Nature in restoring normal action, which constitutes health.

In all electric currents, there is the positive and the negative. The positive corresponds to the alkaline and hot, and the negative to the acid and cool; and all the elements come under one or the other of these conditions.

The Electro-magnetic or *Faradaic Apparatus* gives us two currents, called the *direct* and *induced*. The direct current of this apparatus is as nearly like the current of the galvanic battery as possible, and be subject to the magnet and vibrating armature. In the absence of the galvanic battery, it is a very good substitute.

The induced current is obtained from the fine wire helix, which is entirely insulated from the other, or coarse wire helix, which gives us the direct or primary current. This current is entirely different from the primary, as the primary represents quantity with little intensity, while the induced current represents intensity with but small quantity.

The primary current produces but slight sensation; while the induced, or secondary, produces all degrees of sensation, from the scarcely perceptible to the unbearable. This current is especially

adapted to arouse torpid organs to their normal action, and to relieve painful conditions; by its positive and negative polarities, changing plus and minus conditions to normal or healthy action. The currents in the Kidder machine are regulated by changes in the four posts, marked A, B, C, and D, and by a metallic cylinder that passes over the helix. In other machines, by a metallic rod, or bundle of soft iron wires that slide inside the helix, and the breaking of the current, or vibration of the armature, by a set-screw and spiral-point, slightly touching the platinum disc on the armature or vibrating spring. The two currents on Dr. Kidder's machines have each three variations, and by him are called six currents. A B, A C, and A D, belong to the primary; but A B is the most powerful galvanic: while B C, B D, and C D, are all of the secondary; and each of these has a different intensity.

The metallic connections in all apparatus require to be kept perfectly clean and bright, and the acid not allowed to come in contact with the metal or with the wood. Neither must the quicksilver come in contact with the platina plate or any other metal, as it will spoil the platina, and injure the metal of any other part of the machine.

Instruments, or Electrodes, for Local Treatment.

We have now many instruments for the convenient medical application of the different forms

of electricity, such as instruments for the eye, ear tongue, throat, vagina, uterus, and rectum, besides sponge-handle and wire-brush electrodes. We have also the electrode slipper, and sponge-handles of various kinds and sizes; electro-puncture needles, insulated male and female catheter electrodes, and various cauterizing instruments. Besides these, we use in our practice metallic plates and bands of various shapes and sizes; altogether making a large assortment of appliances, at once elaborate and useful. In addition to these, every operator should be furnished with various kinds of speculums, and whatever his inventive genius may or can devise. A pair of India-rubber gloves are very useful, as they enable the operator to hold the sponges in each hand without feeling the current, which is often of great advantage. We regard this branch of the healing art as yet in its infancy; inventive genius having scarcely begun to exert its powers in this direction. The field is extensive; and the prospects flattering, that, at no distant future, this branch of therapeutics will occupy its true position in all our medical institutions, and its blessings be shared by the suffering everywhere.

In the following pages, the *Galvanic Bath* is frequently spoken of. This we have proved to be an efficient aid in restoring unbalanced conditions to an equilibrium; and nothing within the range of medical knowledge has proved so universally

applicable to the cure of all cutaneous affections when scientifically applied.

Any *galvanic battery* can be used to good advantage in connection with a metallic bath-tub. One electrode can be in connection with the metallic tube at the foot of the bath-tub or elsewhere, and the other applied locally or generally, as the case may require. The body being a better conductor than water, wherever the electrode is applied on the body, the current passes in or out at that point according as the *negative* or *positive* is applied. For local treatment, a sitz-bath answers very well; and for some purposes a hand or foot bath; or the current may be used as are the currents of the electro-magnetic machine, by sponge handles and electrodes. In giving a full or half bath, the time may vary from fifteen minutes to thirty. In many cases, thorough hand-rubbing, and squeezing of the muscles, will aid much in the cure. We have seen wonderful effects produced in secondary syphilis, indolent ulcers, and mercurial diseases, by this means. Dry rubbing must follow the bath; and, in cool weather, the patient should remain in-doors for half an hour or more, to avoid taking cold.

CHAPTER IX.

ELECTRICAL DIAGNOSIS AND MEDICATION.

Electrical Diagnosis.

IN treating upon this subject, it is well to state that no two persons are precisely alike, either in temperament or susceptibility to electric currents; neither is every part of the system equally sensitive to the current in the same person. Consequently, the same strength of current does not produce the same effect on every individual. Fleshy persons feel it less sensibly than lean. This being the case, we must, in our diagnosis, make due allowance for any difference which may exist. As a general rule, those parts where the bones are thinly covered with muscle and fat feel it the most, especially if prominent nerves pass over the bones : hence the forehead and scalp, shoulder-blades, and over the ribs and sternum, shins, hands, and internal ear, are among the most sensitive parts of the body.

When any part of the body is more sensitive to the current than natural, and a dull or sharp pain is produced, we infer that there is a too positive condition of that part or organ; and, on the con-

trary, if there is a lack of natural sensibility, we conclude that organ or part is too inactive or negative. In treatment one will require soothing, the other tonic and stimulating applications.

As the hair, in its dry state, is a non-conductor of electricity, we first, in commencing our diagnosis, moisten the hair; next seat the patient on the negative electrode, and with the sponge electrode in one hand, and the other moistened with water, with a very gentle current, touch the upper part of the spine to test its strength. If right, commence on the top of the head, and pass the hand down on all sides to the neck; lastly, pass the hand over the forehead and upper part of the face; and if any sensitive and painful spots appear, these spots require treatment with the positive electrode; but, if you find any part in the opposite condition, it needs treating with the negative. Sometimes diseased parts may be known by their unnatural heat: such conditions always denote positive action; and sometimes the hand alone is sufficient to allay the irritation.

After thus diagnosing the head, increase the current, and with either two or three fingers, or a small sponge electrode, pass gently down the centre of the spine to its base. Then make similar passes down each side of the spine; and, if no tenderness appears in any part, increase the current, and repeat the operation. If, after doing so with a pretty strong current, no tender

places appear, you may conclude the spine is not at fault, unless some parts are devoid of natural sensation : in that case, there is a lack of action, and may be paralysis; and, in either case, the treatment would have to be according to the case.

After diagnosing the spine, place the positive electrode on the cervical vertebræ, and pass the negative over the scapula, clavicle, and upper part of the sternum and chest; then pass the negative from the spine, under the arms, and over the chest, lowering the positive on the spine to a little above the negative. In this way pass the negative over the kidneys, stomach, liver, spleen, pancreas, and bowels, down to the pubes. Wherever there is more or less sensibility than natural, or an enlarged, contracted, or torpid state, the parts are either unnaturally positive or negative, and need treatment according to their respective conditions.

Next, seat the patient on the positive; and, if a male, pass the negative over the spermatic cords; and, if there is great tenderness or susceptibility to the current, there is indicated seminal weakness, impotency, or some other trouble of that nature, especially if the scrotum is flaccid, and the testicles sensitive to the touch or a light current. If a female, and there is great weakness of the abdominal muscles, and tenderness over the uterus and ovaries, with pain in the sacral and lumbar region, there is prolapsus uteri and leucorrhea,

with dyspepsia and nervous prostration. If there is any enlargement over the ovaries, there may be ovarian tumor. (In order to diagnose the internal organs, it may be necessary to use the speculum).

Next, place the feet on a metallic plate, or large, moist sponge, with the negative, and apply the positive from the roots of the sciatic nerves down to the popliteal spaces between the tendons; thence down each side of the legs, to the inner and outer sides of the feet; and if any unnatural sensitiveness is found in any part, on the whole course of the nerves, treatment is needed: it may be a case of sciatica, and must be treated accordingly. In making your diagnosis, have the parts exposed as little as possible, and *rub every part dry* as you finish its diagnosis. Make the diagnosis as speedily as you can, and avoid producing unpleasant sensations as much as possible, and be sure to avoid giving any shocks. If your patient is very nervous and sensitive, better leave some unimportant part undiagnosed, than either frighten or fatigue your patient.

In addition to the above, it is well to examine the pulse and tongue; and, if the lungs are affected, apply the stethescope or ear. Use all the discrimination and judgment you can to comprehend the real condition of your patients, and gain their confidence; knowing that a correct knowledge of the disease is absolutely essential to the cure.

General Tonic Treatment.

As this treatment is often alluded to in the following pages, we will here speak of it specifically. Seat the patient on the positive electrode, secondary current; and either with the sponge electrode, or the metallic handle in a sponge, treat all the way up the spine to the cervical vertebræ; also over the sides, chest, and abdomen, and over the arms and hands; then moisten the hair, and treat gently the cerebellum, and give a few passes also over the forehead and temples. Finish by placing the positive at the feet, and treat with the negative all up the limbs to the body. Treat from twenty to thirty minutes, and rub each part dry as you finish treating it. In treating all kinds of cases, the strength of current and time required will depend on the nature of the case, and must be left to the judgment of the operator, after making a careful diagnosis. Shorter treatments need to be given to the nervous and much prostrated than to the more robust, as their systems are capable of little re-active power, which consideration should always govern our treatment, as to the length of the séance and strength of current.

MEDICATION.

We are aware that some very intelligent operators entirely discard every kind of *medication*, and

claim that electricity, unaided and alone, is sufficient for all emergencies. To all such we would say, you are welcome to your one-idea system, for such it really is; but it can scarcely claim a relation to the progressive developments of the present age. For ourselves, we can say that we positively know of many remedies that are powerful aids in many cases, where the best electrical treatment alone would either fail, or require a much longer time to effect a cure without them. Our practice will continue to be, not to discard any known and reliable means which we have proved to be efficacious in curing disease, and relieving the sufferings of our fellow-creatures.

Having thoroughly studied and tested the merits of hydropathy, and practiced extensively according to allopathic rule; having also graduated in the homœopathic school,—we have reason to know that each has its merits; and we would advise all students to give their patients the advantage of the best medical treatment within their knowledge, not discarding any known remedy, of whatever school it may be, or from whatever source derived. This do in connection with your electrical treatment. Besides, we would advise every advantage to be taken of all hydropathic and hygienic agencies within our knowledge, human magnetism included: for, be assured, there is healing power in healthy organisms; and, when there is a *will* that *good* should be done to the

patient, it can be done. This is just as certain in this day, as that Jesus opened the eyes of the blind, made the lame to walk, and cleansed the lepers, two thousand years ago. In a word, we say, be free to use all available means for good, and, our word for it, you will heal more sick, and make more wonderful cures, than by limiting yourself to any one alone.

CHAPTER X.

TREATMENT.

Inflammation (from inflammo, to burn).

INFLAMMATION is characterized by heat, pain, redness, attended with more or less tumefaction and fever. It is divided into two species, viz., phlegmonous and erysipelatous, and subdivided into acute and chronic, local and general.

Phlegmonous inflammation is known by its bright redness, tumidity, and proneness to suppurate, and by its heating and pulsatory action. It has three terminations, viz.: *resolution,* when there is a gradual abatement of symptoms; *suppuration,* when the inflammation does not readily yield to appropriate treatment or remedies; and *gangrene,* or *mortification,* when the pain abates, the pulse sinks, and cold perspiration appears.

Acute inflammation runs a rapid course: the pulse is full and bounding; the skin, hot and dry.

Chronic inflammation is milder, and of longer duration. *Local* applies to a part, and *general* to the entire sytem. *Erysipelatous inflammation* is of a dull-red color, superficial, and merely of the

skin; spreading *unequally*, with burning and stinging, and generally ends in vesicles, or desquamation.

CYSTITIS, OR INFLAMMATION OF THE BLADDER.

Symptoms, or Diagnosis.—Violent burning, lancinating or throbbing pain in the region of the bladder, sometimes extending to the perineum, genitals, and upper part of the thighs: the pain is increased by pressure over the pubes and perineum. There is frequent efforts to urinate, but without success, except in drops with severe pain. Bowels are constipated; pulse, full and hard; skin, hot and dry; thirst, urgent; with sickness of the stomach, and vomiting.

Causes.—Mechanical and irritating substances in the bladder; urine retained too long (a frequent cause with ladies); external injuries; cold, suppressed perspiration; hemorrhoidal discharges; cantharides, turpentine, &c., or metastasis of gout and rheumatism.

Treatment.—Place the patient in a shallow, warm sitz bath, with a sponge and negative electrode at the perineum, with moderate induced current; and treat over the kidneys, lumbar region, pubes, and lower part of the abdomen, for ten minutes: if urination takes place, all the better. Then put the feet in a pail of warm water, with the negative electrode, and treat with the positive over the kidneys, spine, abdomen,

pubes, perineum, and genitals, for ten minutes. If a male, place the genitals in a cup, with the negative electrode, having the water warm. Treat in this way for five minutes, applying the positive over the bladder and pubes; let this finish the treatment.

For soothing the nerves, and promoting free urination, a full warm sitz bath, taken for fifteen or twenty minutes before going to bed, is often of great service. The diet must be moderate, and of a cooling and soothing nature: all extremes of heat or cold avoided; the room kept cool and well ventilated. Quiet must be observed, and effort avoided, by the patient.

PHRENITIS, OR INFLAMMATION OF THE BRAIN.

Diagnosis.—High fever, violent headache, redness of the face and eyes, throbbing of the temporal arteries, intolerance of light and sound, watchfulness, and delirium, which often becomes furious.

It commences with a sense of fulness in the head, pulse full, restlessness, disturbed sleep, or its entire loss.

Causes.—Blows, falls, or other injuries of the head; suppressed eruptions, and habitual discharges, and overtaxing the mind.

Treatment.—Seat the patient on the negative, by means of a moist sponge inclosing the electrode: moisten the hair all over the head, and either by means of a metallic cap to fit the head, or large

moist sponge, apply the positive to the head, with a light primary current, but not so strong as to be unpleasant to the feeling, bearing in mind that the head is very sensitive to the current, even in its normal condition. Make steady and gentle passes down the spine from all parts of the head, ceasing for a moment every few minutes. After treating in this manner for ten or fifteen minutes, change to the secondary current, and, with a moderate current, treat from the neck down to the base of the spine for five to ten minutes; then, laying down the electrode, operate gently, making passes with both hands from the forehead all down the spine for five minutes; then have the patient lie down, with the head well elevated, and, if there still continues heat and fulness, apply cold compresses to the head, until the heat and pain is reduced. A bag of pounded ice is often of great use, placed under the back part of the head. Keep the room dark, cool, and quiet; give cold, acid drinks and cooling diet, until the inflammation and fever subside. The electrical applications may be renewed every two hours, until the fever has abated, or the polarities are changed. Warm derivative foot and hip baths also aid in restoring the equilibrium. If the patient is unable to sit up, the electrode may be placed under him between the hips, or at the feet. This rule holds good in all cases; for, if we cannot always do just as we

would like, we must do the best we can under the circumstances.

PNEUMONIA, OR INFLAMMATION OF THE LUNGS.

Diagnosis.—The attack commences with chills and flushes of heat; pain in the chest; difficult perspiration; sense of weight; cough, with frothy expectoration, which changes, as the disease advances, to a bloody sputa; pain aggravated by coughing; face bloated and livid (in consequence of the imperfect change of the blood in the lungs); the pulse full, and skin hot and dry.

Causes.—Checked perspiration, by exposure to cold or sudden changes of temperature; violent exertion in speaking, or blowing on wind instruments; inhaling of obnoxious vapors or irritating particles.

Treatment.—There is no disease that requires more prompt treatment than this. Apply the positive to the spine, from the seventh cervical vertebra, to below the shoulders, either by means of the metallic plate for the back, or a large flat sponge, with the negative attached to the chest-plate, having it touch as much of the surface as possible. Use the secondary current as strong as can be endured without pain. Move both the plates gradually down to below the chest, reversing them every few minutes, so as to bring both front and back under the positive pole. Keep up the treatment until the congestion and pain is

relieved. The same may be repeated every two hours, until an equilibrium is restored, and the patient out of danger. Cooling and mucilaginous drinks may be given, and the chest covered with a moist compress, frequently changed, care being taken to prevent any chilly feeling during the change. Sore or tender places may be treated on general principles; and, to restore the strength, general tonic treatment may be given as soon as the case will warrant. Keep the room cool, quiet, and well ventilated, as *pure* air makes *pure* blood, which in turn restores health and strength. To prevent congestion of the brain keep the head cool.

The above may be varied by seating the patient on the negative, and treating with the positive over every part of the chest, beginning at the apex of the lungs, and treating to below the diaphragm.

PLEURITIS, OR INFLAMMATION OF THE PLEURA.

Diagnosis.—Pleurisy is an inflammation of the lining membrane of the chest. It commences with a sharp, lancinating pain, generally in the right side; increased by a full inspiration; pulse full and hard; skin, hot and dry; and severe pain upon intercostal pressure, *i. e.* on the spaces between the ribs.

Causes.—Over exertion, suppressed perspiration, cold, or exposure.

Treatment.—Treat the painful parts by means of a large sponge electrode, positive pole, and secondary current, but not so strong as to be painful. Seat the patient on the negative, or apply it on the opposite side or in the opposite hand. Continue the treatment until relief is obtained, which will usually be in five or fifteen minutes. Treat twice a day or oftener; and, as the pain and soreness leaves, make the treatment more general, with a view to balance perfectly the electrical polarities of the system. Make cooling applications, as in pneumonia.

GASTRITIS, OR INFLAMMATION OF THE STOMACH.

Diagnosis.—Pain in the stomach, with a burning sensation; loathing of food, retching, vomiting, increased by taking anything warm; hiccough; pulse small and hard; tongue coated in the centre, with the edges and tip red and shining, with fever, anxiety, and mental disquietude.

Causes.—Acrid substances, arsenic, and corrosive sublimate; crude articles of food, unripe fruit, drinking much cold water, taking ice-cream or iced fruits when heated by exercise.

Treatment.—Seat the patient on the negative by means of a sponge, S. C., and apply the positive over the stomach, liver, small and large intestines, also over the kidneys, loins, and lumbar vertebra. Commence with a gentle current, and increase as it can be endured without increased pain. After

treating ten to fifteen minutes, lay down the electrode, and manipulate with the warm hand. If there is constipation, give a warm injection of rice-water, and procure an evacuation. Let the drinks and diet be of a mucilaginous kind. Small bits of ice may be placed on the tongue or swallowed, if agreeable to the patient. Treat three or four times daily. Cool compresses applied to the stomach are often very beneficial.

HEPATITIS, OR INFLAMMATION OF THE LIVER.

Diagnosis.—Pain in the right hypochondrium, extending to the right shoulder; pains much increased by pressure on the part, sometimes cutting, but often dull and obtuse; sometimes there is cough, difficulty of breathing, thirst, loss of appetite, constipation, stools clay-colored, urine high-colored and scanty. If the disease continues a few days, or becomes chronic, the skin becomes yellow, and the eyes also, constituting jaundice: there is languor and proneness to sleep.

Causes.—Contusions and blows, inflammations of other organs, great changes from heat to cold, fevers, suddenly-suppressed bilious diarrheas, gall-stones, mineral poisons, and intoxicating drinks.

Treatment.—Apply the positive over the liver, by means of a metallic plate, with a moist, folded towel under the plate, covering the most tender part. Apply the negative by means of a large sponge-

handle over the left side, front, and back, passing the negative down to the base of the spine, and also over the lower part of the abdomen. In severe cases, give two or three treatments a day, until relief is obtained; then once a day till well. Keep the bowels free and regular; diet, &c., as in gastritis.

For Enlarged Liver.—Treat with the negative over the liver, and positive on the opposite side and spine, for five minutes; then treat the liver with positive S. C. as strong as can be well endured; negative on left side and spine, down to the coccyx; then over the spleen and bowels, being careful to follow their course from the duodenum to the descending colon.

For Torpid and Hardened Liver.—Positive on the spine, from the neck downwards; negative over the liver ten minutes; then positive on the liver and bowels, and negative at the coccyx ten minutes: treat daily with strong current.

For Biliary Calculi.—Treat over the liver (and gall-duct in particular) with negative *hot* sponge, and strong current, so as to expand the parts; positive on the spine and back part of the liver. Do this for fifteen minutes; then seat on the negative, and apply the positive over the whole liver, beginning on the back, and finish over the gall-duct where it empties into the duodenum. As soon as the calculi have passed, the patient will feel relieved. Treat two or three times daily

until relief is obtained, and afterwards give general tonic treatment, not neglecting the liver and spleen. By examining the fœces, the gall-stones may be found, if not too large to pass through the duct. Hot fomentations are often of great service.

ENTERITIS, OR INFLAMMATION OF THE INTESTINES.

Diagnosis.—Burning or aching pain about the umbilicus; bowels generally sore to the touch; great constipation, except when the inflammation is in the inferior portion of the intestinal canal, when dysenteric evacuations occur. There is urgent thirst, nausea, and vomiting; skin hot and dry, tongue dry and furred, with edges pale; urine high-colored, scanty, and passed with difficulty.

Causes.—Hardened fœces, colic, spasms, intussusception, drastic purgatives, hernia, mechanical injuries, worms, and metastasis from erysipelas, gout, and rheumatism.

Treatment.—Seat the patient on the negative S. C., and, either with a large sponge or the chestplate, treat all over the abdomen with positive for ten or fifteen minutes. If there is constipation, give warm mucilaginous injections, and repeat the electrical applications every two or three hours, until the symptoms improve. Frequently there is great relief from the first treatment. Sometimes warm fomentations between the treatments are of great service. Diet, &c., as in gastritis.

NEPHRITIS, OR INFLAMMATION OF THE KIDNEYS.

Diagnosis.—Pain in the region of the kidneys, shooting along the course of the ureter (drawing up the testes, if a male), numbness of the thigh, vomiting; urine high-colored and voided frequently, but with difficulty; pulse full, hard, and frequent, in the early stages; later it is small and more frequent. The disease runs a rapid course, and seldom continues beyond the seventh day.

Causes.—Contusions, strains of the back, colds, violent exercise, irritating articles, such as cantharides, turpentine, savine, and metastasis from gout, and rheumatism.

Treatment.—Apply the positive over the kidneys and along the course of the ureters to the bladder; S. C., seating the patient on the negative for ten minutes; then place the genitals in the cup, with the negative, for five minutes; next place the negative at the feet in warm water, and treat with the positive over the kidneys, ureters, and bladder for five minutes. Treat twice or three times a day until the pain ceases, then once a day till well. Drinks, diet, &c., as in cystitis.

SPLENITIS, OR INFLAMMATION OF THE SPLEEN.

Diagnosis.—Fever, pain in the left side, heat, tumor, pain increased by pressure, but not very acute, eyes and skin slightly yellow, and urine tinged with bile, burning in the stomach, vertigo

on rising up in bed, dyspetic symptoms, and fretfulness.

Causes.—Long-continued intermittent and remittent fevers, the abuse of quinine, marshy effluvia; and sympathetically from inflammation of the stomach.

Treatment.—Seat the patient on the negative S. C., and treat over the spleen, stomach, liver, and intestines with the positive ten to fifteen minutes, twice daily. Warm fomentations aid in the cure. Keep the bowels free. Diet, &c., as in inflammation of the stomach.

DIAPHRAGMITIS, OR INFLAMMATION OF THE DIAPHRAGM.

Diagnosis.—Alternate chills and flushes of heat; pain on the diaphragm in coughing or sneezing, or pressing on the chest and abdomen; pulse full, with cough, and sometimes delirium.

Causes.—Same as in pleurisy.

Treatment.—Seat the patient on the negative S. C., and apply the metallic girdle and pad over the diaphragm for ten minutes; then use sponge-handle from the diaphragm to the lower part of the body and spine for five minutes. Treat two or three times a day until the pain ceases, then as the case seems to require. Diet, &c., as in gastritis. Fomentations of warm water are good in this disease.

OPHTHALMIA, OR INFLAMMATION OF THE EYE.

Diagnosis.—Redness, congestion of the membranes, pain, intolerance of light, profuse lachrymation, sometimes a purulent discharge, sensation as if sand were in the eye; when deep-seated, severe headache is often a constant attendant.

Causes.—Exposure to cold winds, suppressed habitual discharges, vivid rays of the sun, close study, mechanical injuries, syphilis, or gonorrheal matter applied to the eye.

Treatment.—The eye being a very delicate organ, great care must be taken not to have the current too strong. First moisten the hair from the crown to the neck, fill the eye-glass with pure soft water, attach the positive S. C., and place the cup to the eye so that it may all be in contact with the water; apply the negative to the back of the head and down the spine for five minutes, then change the water, and proceed as before for fifteen or twenty minutes. Give two treatments daily, and other local or general treatment if necessary. Keep the room dark, but well ventilated; avoid rubbing the eyes, and finish the treatment by making gentle passes with both hands from the forehead backward down the spine. The diet must be light and cooling, and promote a free general circulation, keeping the extremities warm, and the bowels free and regular.

THICKENED OR GRANULATED EYELIDS, AND STYE.

Treatment as in the preceding inflammation, except by first using P. C., and finish with the secondary. If there is much debility, give general tonic treatment; and, if the disease is from scrofulous or other impurities in the system, treat accordingly.

CATARACT.

This is a species of blindness, arising from an opacity of the crystalline lens or its capsule, which prevents the rays of light passing to the retina. The first symptoms are as if particles of dust were in the eye, or floating before it. At first there is no change in the appearance of the eye; it is first turbid, then grayish, ash color, or white: it is sometimes hard, soft, or fluid, or like curd.

Causes.—Often obscure; mechanical; deep-seated inflammation; exposure to intense light. It is often hereditary from a scrofulous diathesis, or syphilitic taint.

Treatment.—By means of the eye-cup, apply the positive P. C. to the eye; negative on the cerebellum, neck, and spine. Have the current as strong as can be endured without pain: remove and replace the sponge every half-minute; continue for ten minutes each sitting. Treat once a day for a week or ten days, and, if no improvement takes place, cease the treatment, and resort

to the electrical machine. Place the patient on the insulated stool, and draw sparks from the opaque part of the eye, having the lid closed. This may be done until some redness appears; then cease, and repeat next day, or more seldom, as the case seems to require. Nice discrimination is needed in such cases, for it will not be well to produce much irritation. The eye must be entirely shaded from the light during the rest of the day, and all excitement avoided, and perfect quiet observed. Diet light and cooling, and rooms well ventilated. If the system is reduced, give general tonic treatment: if there is no change for the better in twenty days, the case is very doubtful.

AMAUROSIS.

Diagnosis.—Paralysis of the optic nerve, with a want of contraction and dilatation of the pupil, which sometimes appears dull or glossy.

Causes.—Organic diseases of the eye, but often from turgescence on some part of the eye; contusions, over-exertion, convulsions, pregnancy.

Treatment.—Negative S. C., with the eye-glass to the eye; and, after moistening the hair from the crown to the neck, apply the positive over the back of the head and neck, for three minutes; then cease for a few seconds, and treat again for ten to fifteen minutes, changing the eye-glass from one eye to the other every five minutes. If the causes remain, try and remove them, and give

general tonic treatment if needed. These chronic diseases sometimes require lengthy treatment to effect a cure ; but the improvement is soon evident.

FISTULA LACHRYMALIS, OR OBSTRUCTION OF THE LACHRYMAL DUCT.

Treatment.—Apply, by means of a small sponge electrode, the positive to the gland over the lachrymal duct, S. C., and, with a small electrode inside the nostril, apply the negative ; but, if the current is too strong, treat with the P. C. A little general treatment, especially about the head, eyes, face, and throat, is very good. Give treatment daily, until the duct is free and healed.

STRABISMUS, OR SQUINTING,

When congenital, may require a surgical operation; when not congenital, it is often readily cured.

Treatment.—Apply the positive P. C. to the relaxed muscles, and negative to the contracted ones, for a few moments at a time, repeating the same from five to ten minutes. Repeat the treatment daily, until restored.

MYOPIA, OR NEAR-SIGHTEDNESS.

When congenital, it is doubtful of cure ; but when caused from congestion of the humors of the eye, and abuse of mercury, it may be cured.

Treatment.—Apply the positive S. C. by the eye-glass to the eyes alternately with the negative on the spine and back of the head; and, to counteract the mercury, seat the patient on the positive, and treat all over the body with the negative P. C. This may be alternated with the treatment on the eyes. Favorable results may be expected within a week. The eyes must not be taxed much during treatment: all the evacuations must be kept natural; food nutritious, but of easy digestion.

PRESBYOPIA, OR FAR-SIGHTEDNESS.

This depends on a flattened state of the cornea, or the crystalline lens, caused by mercury and ardent spirits. The treatment is the reverse of myopia, except the mercurial part, which is the same. This eye-treatment is good for elderly people, and for all weak conditions of the eyes.

PERITONITIS, OR INFLAMMATION OF THE PERITONEUM.

Treatment.—Same as inflammation of the bowels.

OTITIS, OR INFLAMMATION OF THE INTERNAL EAR.

Diagnosis.—Throbbing and excruciating pain in the ear, frequently extending through the head, with fever.

Causes.—Cold, and a result of measles, scarlet fever, and other eruptive diseases.

Treatment.—In the acute and early stage, treat first with the positive S. C., by means of the sponge electrode, over the painful ear or ears, negative in the hands; then negative on the spine, between the shoulders and neck. Next apply positive, by means of ear electrode, inside the ear, passing the negative from the neck down the spine, for ten to fifteen minutes altogether. Do not have the current so strong as to increase the pain. If the ears have commenced to suppurate, use the P. C., and reverse the treatment by treating the ears with the negative, and occasionally placing the positive on the tongue. Keep the ears as clean as possible, by syringing them gently with tepid castile soap-suds, twice a day. Treat once or twice daily, as the case seems to require. For further treatment of diseases of the ears, see "Deafness."

LARYNGITIS, OR INFLAMMATION OF THE LARYNX.

Diagnosis.—This disease commences with symptoms common to other inflammatory affections, such as chills, alternating flushes of heat, soreness in the fauces, uneasiness in swallowing, voice hoarse, breathing laborious, pulse full; face flushed, sometimes purple; eyes staring, evincing great suffering.

Causes.—Cold, suddenly suppressed perspiration.

Treatment.—Place the positive S. C. on the

tongue, and with the negative treat the throat, neck, and chest; finish by seating the patient on the negative, and treat the neck and throat and upper part of the chest with the positive, all occupying ten to twenty minutes. Treat twice or oftener daily, as the symptoms indicate. This disease runs a speedy course, and needs careful and prompt treatment. Cooling compresses and acidulated drinks and gargles are often very grateful. For food, thin oatmeal gruel, arrowroot, rice or barley, soft toast, or soaked butter crackers. In all inflammatory and febrile diseases, all the evacuations must be kept free and natural.

TONSILITIS, OR INFLAMMATION OF THE TONSILS, (QUINSY).

Diagnosis.—Throat sore, swallowing and breathing difficult, cutting pains, tonsils red and swollen; throat dry and parched, with much mucus, hard to be detached; pulse accelerated, tongue foul, breath offensive.

Causes.—As in laryngitis.

Treatment.—Apply positive P. C. with tongue and throat electrodes on the tongue and to the swollen parts of the throat, and negative by means of metallic band around the body, over the stomach, liver, and spleen. Then apply the positive by means of a sponge electrode over the throat, neck, and upper part of the chest, all occupying fifteen to twenty minutes. Treat as often as the

symptoms indicate. Diet, drinks, &c., as in laryngitis.

GLOSSITIS, OR INFLAMMATION OF THE TONGUE.

Diagnosis.—Tongue hot, dry, and swollen : it often entirely fills the mouth.

Causes.—Calomel, mechanical injuries, sting of insects, cold.

Treatment.—Positive S. C., by tongue electrode, and seated on the negative; current good strength, for ten minutes; then with sponge electrode treat the neck, ears, throat, and chest for ten minutes; treat often until the swelling abates. Drink, diet, &c., as in laryngitis.

TRACHITIS, DIPHTHERIA, AND CROUP.

Diagnosis.—Symptoms of cold; hoarseness when crying; cough hoarse, with hollow sound, and crowing noise; next shrill, breathing more difficult, face flushed and swollen, eyes protrude on coughing, the head is thrown back, with anxious expression of countenance. This disease requires the closest watching and most energetic treatment.

Causes.—Cold is the most frequent; sleeping in a current of air, especially during perspiration.

Treatment.—Seat the patient on the negative P. C., and treat the throat with positive from the chin to the chest; then treat the sides of the neck, from the ears to the chest; and, if practicable,

introduce the throat or tongue electrode as near the roots of the tongue as possible. If the throat electrode is used, fasten a small piece of soft sponge, *moistened*, on the ball, and apply for three minutes, and repeat several times. Finish by a general treatment with the positive, from the throat down to the stomach, and base of the spine. The whole treatment should be immediately repeated if the danger has not disappeared; applying the negative on the neck and cerebellum, with positive in front and on the tongue; *no time must be lost*. For drinks, give toast-water, or weak lemonade; avoid all mucilaginous drinks. Keep the patient as quiet as possible, and the room well ventilated. Diphtheria is treated as above, with current strong as can be well endured. Cooling compresses may also be applied.

BRONCHITIS, OR INFLAMMATION OF THE BRONCHIA.

Diagnosis.—Commences like a common catarrh, with slight cough, and sense of tightness and oppression in the chest. At first the cough is dry, but soon followed by copious secretion of tough mucus, and abatement of the cough. The acute form of bronchitis much resembles croup in children.

Causes.—Same as in general inflammation of the organs of the chest and throat.

Treatment.—Apply the positive on the tongue, as near the root as possible, light S. C., and nega-

tive on the spine below the shoulders, and also over the stomach and bowels for eight minutes; then, with sponge electrode, treat with the positive over the chest, from the neck down to the diaphragm, keeping the negative at the same time a little lower on the spine for ten minutes; then seat the patient on the negative, and treat with the positive, all the way from the neck to the abdomen and lumbar vertebra. Treat once or twice daily, until the symptoms improve; then every other day until well.

Diet and regimen, same as in trachitis.

CHRONIC BRONCHITIS.

Diagnosis.—Uneasiness in the chest; rattling, or a loud respiratory murmur; cough, hoarseness increased by talking; hawking and scraping of the throat; the expectoration varies; pulse small and frequent; palms of the hands hot, with mental and physical irritability.

Causes.—Over-exertion of the voice, constitutional predisposition, asthma and catarrh, colds and exposure.

Treatment.—So long as there is irritation and soreness in the parts, same as in inflammation of the bronchia; but, when that and the cough cease, first treat with the positive S. C., on the spine, and negative on the chest, moving both downward to the abdomen and hips at the same time, for ten minutes. Then seat on the positive, and treat all

over up to the cerebellum with the negative, to tone up the whole system, and restore an equilibrium of all the forces.

Let the food be nutritious and of easy digestion; go out daily, but do not exercise so as to feel much fatigue; be cheerful and hopeful as possible.

INFLAMMATION OF THE PALATE.

Diagnosis.—Palate much inflamed, red, swollen, and elongated, resting on the tongue, with burning sensation, and constant inclination to swallow, attended with some difficulty.

Causes.— Cold, and injuries from foreign substances; when it accompanies inflammation of the throat and tonsils, it is only symptomatic.

Treatment.—Attach a soft, moist sponge to the throat electrode, and apply to the palate, positive P. C., and negative on the back of the head and neck, and down the spine. Next place the positive at the root of the nose, negative on the neck, and down the spine as before. Treat twice daily, for fifteen or twenty minutes, and use cooling drinks, avoiding extremes of heat and cold.

INFLAMMATION OF THE GUMS.

Diagnosis.—Gums red, swollen, spongy, easily bleed, and are covered with small white canker-sores; mouth hot and dry; tongue coated, foul, and slimy; breath foul, with putrid taste.

Causes.—Dental irritation and decayed teeth.

Treatment.—As for glossitis, or inflammation of the tongue.

PARATITIS, OR MUMPS.

Diagnosis.—Some febrile excitement, stiffness in the jaws, swelling of one or both of the parotid glands, increasing until the fourth day, when the cheek is much swollen, firm and tender; on the fifth day the swelling begins to subside, perspiration sets in, and the urine deposits a red sediment. To prevent metastasis, keep the face warm, and not take cold. This disease is somewhat contagious.

Treatment.—Place the metallic band around the waist, with negative S. C., and with the sponge electrode positive treat freely over the swollen parts, and down the spine. Next put the feet in warm water with the negative, and treat with the positive from the swollen parts to the base of the spine, altogether occupying thirty minutes. Treat twice a day until well. Diet light and cooling; keep the feet warm and avoid taking cold.

INFLAMMATORY RHEUMATISM.

Diagnosis.—Pains in the joints, swelling and redness; the perspiration is of a urinous odor; urine albuminous, and diminished in quantity; it is often preceded by gastric derangement.

Causes.—Exposure to cold or wet in changeable weather, dissipation, abuse of mercury. It some-

times follows scarlet fever, measles, dysentery, and suppressed habitual discharges.

Treatment.—If in the shoulders or arms, take the sponge electrode in one hand, and sponge handle in the other; place the positive on the lame part, with the negative a little lower, and pass the current through and through, passing down to the hand; then place the hand with the negative in warm water, and treat with the positive from the spine down to the hand; use the S. C. If the spine is affected, seat on the negative and treat with the positive to the base of the spine; the current strong as can be endured without much increase of pain. If the disease is located about the heart or stomach, treat with positive, using a light current, with gentle, even pressure over these organs. If the hips, thighs, knees, or feet are affected, treat through and through as on the arms; then place the negative in water, placing the feet in it, and treat freely with the positive all the way down. After the disease is cured, give a few general tonic treatments, to establish the general health and tone of the system.

CHRONIC RHEUMATISM.

Diagnosis.—Enlarged and stiff joints, hands and feet often much deformed, and occasionally painful.

Treatment.—To reduce the enlargement, first

use the negative P. C. over the parts (the pure galvanic is best), positive on the opposite side, but a little lower on the arm or leg; current moderate; then treat with the positive on the enlarged part, S. C., with the hand or foot in water, as in acute rheumatism. All the other parts are to be treated on the same principle, with occasional general tonic treatment; freely rubbing with the hands aids in the cure. Relief may be expected after the fourth treatment. Sometimes the symptoms are worse at first, but soon improve. Courage and perseverance are very necessary in these old chronic cases. Electro-chemical baths are excellent in both acute and chronic cases, when available.

ARTHRITIS, OR GOUT.

This affection is commonly confined to the joints, generally the great toe. It is a positive or acute disease, and requires to be treated precisely as inflammatory rheumatism of the joints. We must find the cause and remove it, or our cure will only be temporary.

GONITIS, OR INFLAMMATION OF THE KNEE.

Treatment.—Whether from injuries or rheumatism, is treated as rheumatism.

In case of *white swelling*, which is often caused by a scrofulous taint in the system, general tonic

treatment should precede the local. Use the P. C., treat locally through and through the joint; then place the negative at the foot, and treat with positive from the knee downward. This being a chronic affection, requires time to cure, if it is curable, which sometimes is not the case.

HIP DISEASE.

This is to be treated like white swelling, even when there is a discharge. Wash the sore with castile soap and tepid water.

FEVER.

In a normal condition of the system, the electrical or vital currents flow from the centre of vitality to the periphery or surface; and, whether this condition is disturbed by local or general causes, fever, general or local, is liable to follow; and the manifestation may be constant, intermittent, remittent, typhoid, or eruptive.

If the cause is local, we must treat it according to its positive or negative symptoms, with a view to restore normal action, or balance the electrical polarities of the affected parts.

When there is general feverishness all over the body, with skin hot and dry, attended with headache, place the feet in warm water with the negative S. C., and, after moistening the hair, commence on the head with a very light current, using the open hand as the positive electrode, moistened with

water. After a few minutes' treatment of the head, increase the current, and with a warm, moist, large sponge electrode, treat freely for fifteen minutes all the way to the feet, being careful to rub dry as you finish each part. Keep all the surface warm.

After finishing with the electrode, rub all over with the dry, warm hand, and cover up the patient in bed. Repeat the treatment as often as the fever returns.

In remittent and intermittent fevers, during the cold stage, give opposite treatment to the above, omitting the head, and paying more especial attention to the spine; for that is the great nervous highway of the system, from which all the parts and organs receive their nerve-power. By this treatment, all fevers may be cut short in their action, and fever and ague cured within a week,— often in one or two days.

In eruptive fevers, embracing scarlet fever, measles, small-pox, chicken-pox, &c., the object must be to aid nature in throwing the morbid elements to the surface, thus relieving the internal organs and system generally. Next, to carry off the superfluous heat and fever. The first is done by placing the positive at the feet, and treating generally with the negative; the second by reversing this treatment. Finish the treatment by the last mode, and all will be well.

In treating those parts where the eruption has come to the surface and is discharging, and all

sores and ulcers, we cover the parts with a moist cloth, wrung out of warm water, and treat over the cloth. This prevents the matter from getting on the sponge, and is more pleasant to the patient. When the surface is sore or tender to the touch, the rubbing motion of the electrode should be avoided. Keep the patient quiet, and the room of even temperature but well ventilated; change the linen frequently. Great care is needed in relation to diet; let the food be light and cooling, even when convalescent. In all fevers, the general electro-magnetic bath, is one of the most direct means of subduing the disease.

In treating for fever and ague, the liver and spleen require special attention, particularly when either or both are enlarged, or the liver torpid. Remember, that where you place the positive, it induces contraction, and lessens engorgement and soreness; while the negative increases action, induces relaxation and the elimination of morbid elements.

CONGESTION.

General Congestion depends on a fulness of habit in those who indulge the appetite, without a proper regard to exercise and healthful evacuations.

Local Congestion results from a want of action, or equilibrium of the circulating fluids; from local irritation, or mechanical pressure; bruises, falls; and mental emotion, such as joy, grief, terror, or fear.

Treatment.—For general congestion, first regulate the appetite and evacuations. Seat the patient on the negative S. C., and treat freely with the positive, from the head to the base of the spine, paying special attention to the kidneys, liver, and spleen. Treat daily, as long as necessary.

For local congestion, treat the congested part or organ with the positive S. C., placing the negative on the negative part, if such is apparent; but, if such cannot be detected, treat with the negative in the most convenient and appropriate part. In treating the head, the hair must first be moistened, and the current light, so as to feel pleasant and soothing.

CHOLERA-MORBUS.

Diagnosis.—Commences suddenly, with sickness of the stomach, and vomiting; severe griping pains, accompanied with purging, and sometimes violent cramps: at first the discharge is a little bilious; but toward the last, the discharge is thin, bilious, and slimy, especially if accompanied with tenesmus.

Causes.—Errors in diet; the heat of summer, particularly when the nights are cool, chilly, or humid; drinking acidulated liquids, or too much iced water when heated; suppressed habitual discharges, or repelled cutaneous eruptions.

Treatment.—Seat the patient on the positive S. C., and treat freely with the negative, up the spine

and over the stomach, liver, and bowels, following the course of the intestines, to the lower portion of the descending colon. If food is taken, let it be partially of milk, boiled in some way. Keep the patient quiet, and in a recumbent position. Treat several times a day if the case seems to need it. Keep all the parts warm, and prevent cramps by treating the parts attacked both by the hands and the current.

CHOLERA, ASIATIC.

Diagnosis.—Sometimes it commences with diarrhœa; at other times with vomiting, and spasms of the muscles of the chest; and frequently with vomiting and purging of a substance resembling rice-water, with cramps in the muscles of the chest, abdomen, and extremities, which almost raises them in knots, with tearing sensations; the surface blue, extremities cold and shrunken; a cold, clammy perspiration, and burning thirst; yet the consciousness is generally retained.

Causes.—Most probably a want of balance, or equilibrium of the electrical currents of the atmosphere, or some unknown agent, which has so far eluded medical research. But the predisposing causes are numerous; among which are improper food and drinks, grief, terror, fear of an attack, excesses of every kind, exposure to night air, &c.

Treatment.—In a normal condition of the system, the currents are from the centres of vitality

to the surface, thereby keeping the extremities warm; but, in cholera, we find they are reversed or unbalanced. Our first object is to restore the lost equilibrium, or change the course of the currents, which are now set inward, to their normal outward flow, and thus correct the deranged action of the stomach and all the other viscera. Hence we place the positive at the base of the spine, and with the negative S. C. treat first up the spine to the cerebellum, and afterward from the spine, all over the chest and abdomen, paying special attention to the stomach, liver, spleen, pancreas, kidneys, and small and large intestines. Commence with a moderate current, and gradually increase it as long as it can be endured with some degree of comfort; next treat the hands, arms, feet, and limbs, in the mean time keeping all the parts warm, and, if need be, do so by means of warm applications; the vitality must be kept up, and the treatment repeated every hour, if necessary, until there is an abatement of the symptoms; free hand-rubbing is very good in helping the circulation. During convalescence, the diet should be mild: at first, simple broths, beef-tea, rice, and toast. Great care is necessary not to exercise too much, or do anything to cause a relapse.

CHOLERA INFANTUM.

This disease is to be treated on the same principle as cholera morbus; but the current must be

light, and adapted to the tender sensibilities of the patient.

DIARRHŒA, ACUTE.

Diagnosis and symptoms very much resemble cholera morbus: the causes also are very similar.

Treatment.—Same as in cholera morbus, with such variations as may seem best to the operator.

DIARRHŒA, CHRONIC.

Treatment.—Give general tonic treatment with S. C., then with the positive P. C., at the base of the spine, by means of moist sponge, or rectum electrode. Treat with negative over the liver, stomach, spleen, and bowels, carrying it up to the shoulder-blades. Treat twenty or thirty minutes, once a day. It may require several weeks to make a cure. Diet nutritious and soothing, and rest in a recumbent position.

DYSPEPSIA.

Diagnosis.—Pain in the epigastrium, wind in the stomach after eating, nausea, acid or bitter eructations, heartburn, costiveness, depression of spirits, languor, headache, nervous prostration.

Causes.—Irregularity and excess in diet, indigestible food, stimulating drinks, sedentary habits;

using tobacco, cathartic medicines, and tonic bitters; irregular hours, and all excesses.

Treatment.—First remove or avoid all the causes of the disease. Give general tonic treatment for several days; then alternate with local treatment by placing the positive P. C. on the neck over the pneumogastric nerve, and treat with the negative over the stomach, liver, spleen, and bowels, from the duodenum to the lower part of the descending colon; let the current be moderate, and the pressure of the electrode even. Next bring the positive over the spine opposite the stomach, and repeat the treatment over the organs and bowels. Let the whole treatment occupy twenty minutes, and give one treatment daily, until the symptoms improve; then every other day. Let the diet be soothing and nutritious, and the quantity moderate, and times of eating regular; exercise moderately in the fresh air, and be free from all care. If the bowels are constipated, finish the treatment by the positive over the bowels, and the negative at the rectum, either with a sponge or rectum electrode. Sometimes a tepid injection of castile soap-suds helps the treatment. Considerable tact is needed to treat chronic cases successfully, but acute cases soon yield to the treatment. Patients often need much encouragement and assurance of a cure; for very often they have tried, as they say, "everything," before seeing you.

CONSTIPATION, OR COSTIVENESS.

Causes.—Inactive liver, neglect of the calls of nature. In pregnancy, pressure of the uterus on the intestines, and many causes such as produce dyspepsia.

Treatment.—Begin with positive S. C. at the rectum, and negative over the liver, spleen, and stomach, for ten minutes; then reverse the treatment for ten more. The first increases the action of the organs; the second promotes the opening of the bowels, and the removing of obstructions. Chronic cases need perseverance. Daily tepid sitz baths and a moist compress over the bowels all night aid in the cure.

DYSENTERY.

Treatment.—Same as for cholera morbus, except when there is constipation: in that case, finish the treatment with the negative at the base of the spine, with a moderate current, for five minutes, and positive chiefly over the large intestines, especially the descending colon. Diet and regimen the same.

COLICA, OR COLIC.

Treatment.—If the bowels are constipated, procure an evacuation by a tepid injection of water. Then seat the patient on the negative, and treat freely with the positive over the small and large

intestines, with a gentle pressure all the way from the stomach, following the course of the intestines to the lower part of the descending colon. Apply the sponge electrode as warm as can be comfortably endured, and continue until the pain ceases, or for fifteen minutes at a time. All kinds of colic are treated the same except the bilious, which requires more treatment over the liver. Hot fomentations are also of great service.

ICTERUS, OR JAUNDICE.

Diagnosis.—Yellow skin, clay-colored fæces, highly-colored urine, loss of appetite, languor, constipation, dull pain over the liver, induration or organic disease.

Causes.—Obstructions in the bile ducts, pregnancy, the too free use of quinine and arsenic in intermittent fevers, ardent spirits, and all other preventives to the free and natural action of the liver.

Treatment.—If the liver is enlarged and hardened, and not painful on pressure, apply the negative S. C. over the liver, spleen, pancreas, and small and large intestines, with positive at the coccix, with moderate current, for fifteen minutes. But, if the liver is sore and painful, reverse the treatment; and, if constipated, promote action by tepid injections. Treat once or twice daily until the liver is restored to its normal condition.

BURNS AND SCALDS.

Treatment.—Same as in inflammation in any local part; only the parts must be covered with a linen cloth saturated in olive oil, and the pressure light as possible. Positive on the inflamed part.

Frost-bitten Limbs and Chilblains are to be treated in the same manner; also all kinds of wounds and bruises. When cloths or compresses are needed, use water instead of oil.

CATARRH.

Diagnosis.—In acute cases, there is irritation or inflammation of the mucous membrane of the frontal sinuses and nostrils, loss of smell, and watery discharge from the nose and eyes. In old chronic catarrh, there is pain in the forehead and between the eyes; the discharge is very offensive, and often passes down the posterior nostrils to the throat, causing a constant scraping, hawking, and gagging, with expectoration, which is sometimes swallowed during sleep, causing dyspepsia, and producing bronchitis from irritation of the trachea and bronchial tubes.

Treatment.—Catarrh in the *acute* form is to be treated by immersing the upper part of the nose and forehead in rather warm water, by means of a wash-bowl, and a sponge of sufficient size to rest the part upon in the water. Place the positive S. C. in the water under the sponge, and with the

negative treat the neck and spine : if convenient,
let the patient hold the negative in both hands.
After treating in this way for five to ten minutes,
seat the patient on the negative, and treat with
the hand over the nose and forehead to the spine,
for five to ten minutes more; and, if there are
symptoms as if the system generally was suffering
from a cold, give general treatment over the whole
body, down to the abdomen and base of the spine,
for ten minutes more ; repeat daily, or oftener,
according to the symptoms.

Common Colds have to be treated as acute catarrh,
except when the surface of the body is chilly. In
that case, seat the patient on the positive, and treat
with the negative up the spine to the cerebellum, and
over the body generally from the spine all around,
until a general feeling of warmth is felt all over
the body. This establishes the natural flow of the
life-currents from the centre to the surface. The
electro-magnetic bath is of great service in all
cases of colds, treating all over the body with the
negative electrode.

Influenza.—I will here say, that influenza, from
whatever cause, is to be treated as acute catarrh
and common colds.

Chronic Catarrh.—Often results from neglect of
proper treatment of the acute form, and sometimes
is the cause of other affections, such as bronchitis,
dyspepsia, and even consumption.

This form of disease requires opposite treatment

from that of the acute, so far as running the currents is concerned, viz., the negative must be applied to the parts affected, and for a little longer time, and the primary current; and, in cases of very long standing, the pure galvanic, alternated with the electro-magnetic, is to be preferred. I have known cases to be cured in this way, that had existed from infancy; but it frequently requires months of steady treatment to make a perfect cure: other cases yield in from one to six months, but all are improved from the beginning.

APHONIA, OR LOSS OF VOICE.

Causes.—Cold, a disease of the trachea; a tumor of the fauces, or about the glottis; paralysis, or suppressed eruptions.

Treatment.—If from paralysis, moisten the hair on the back of the head, and treat with the positive S. C. on the cerebellum, negative on the tongue, and with the throat electrode for five minutes; then with negative over the throat, and from the ears down to the sternum, and over the respiratory muscles of the chest and abdomen; at the same time bringing the positive down the spine, nearly opposite to the negative in front. If there is general debility, give general tonic treatment two or three times a week, and the other treatment once a day, for several weeks; then, if there is much improvement, not quite so often, but sometimes the voice is restored almost imme-

diately. If the disease arises from other causes, treat according to the positive or negative symptoms. For suppressed eruptions, give general tonic treatment: and a general electro-galvanic bath.

PHTHISIS PULMONALIS, OR CONSUMPTION OF THE LUNGS.

Diagnosis.—Prominent symptoms: emaciation, debility, cough, hectic fever, loss of appetite, short breathing, quickened pulse, purulent expectoration, blueish color around the nails, swollen feet and limbs, indigestion, and often a too relaxed or constipated state of the bowels. Different parts of the lungs may be hepatized or hardened, or hollow from the breaking down and discharge of tubercles. Both conditions are readily determined by percussion and auscultation.

If, on inquiry, we find that the disease is hereditary, and it has passed into the second stage, we need not expect to make a cure; but if it is not hereditary, and we commence before that period, we may hope to cure it. We must try to ascertain the cause or causes, and, as far as possible, remove them. Sometimes we need to bring to bear all the resources of our system, with a view to eliminate from the body the morbid elements, balance the circulation, and give tone to the whole nervous system, and through that to all the organs, arresting the further deposit in the lungs, and promot-

ing the absorption of the tubercular matter already there.

Treatment. — When the system is emaciated, we commence by giving general tonic treatment, for a few days or a week, daily. Then after diagnosing, if we find very tender places on different parts of the chest, sides, or back, treat these places with the positive S. C., very gentle current, placing the negative on the spine, a little below the positive. When the soreness is removed, change to the P. C., and with the plates, one on the spine and the other on the chest, with moist cloth under each, connect the back with the positive, and chest with the negative moderate current, and gradually move them down to the abdomen and lumbar region. Repeat this for fifteen minutes, and rub the whole surface with the warm hands. If there is soreness in the throat use the throat electrode with positive, and treat the neck and chest, to the diaphragm, with the negative for five minutes. To improve the circulation in the extremities, put the feet in warm water, with the negative S. C., and treat with the positive from the chest to the feet; next place the negative in the hands, and treat the spine, shoulders, and arms with the positive, each occupying five to eight minutes. All of this may at first be repeated daily,—one part in the morning, the other in the afternoon or evening. The foregoing has special reference to relieving pain, promoting absorption of morbid elements, purifying the

fluids (the blood in particular), balancing the circulation, and giving tone and vigor to the whole system. But when there is a scrofulous diathesis, and a general depravity of the fluids, with cutaneous affections, we resort to the pure galvanic treatment, in connection with the foregoing. This consists in placing the patient in a warm, full bath, and, with a battery of from six to twelve cups, treat, by placing the positive at the feet, or base of the spine, and with negative treat over the whole surface of the body and limbs for ten to twenty minutes. Have the room warm, and rub the patient freely, from head to feet, till the skin is red. The patient must be careful not to go into a cool atmosphere for thirty or forty minutes. This treatment opens all the pores, and is for the purpose of eliminating from the system all morbid elements; thus relieving all the organs, and at the same time restoring the equilibrium of all the vital powers and circulation.

HYDROPS, OR DROPSY.

Dropsy is generally but a mere symptom of some other affection. Its proximate cause consists in an inflammation, congestion, or exalted action of the capillary extremities of the arterial vessels of the serous and cellular membranes, and a torpor or inactivity of the venous absorbents of the same parts. It may be acute or chronic; caused by loss of blood and other fluids; general debility

resulting from disease; mechanical injuries; obstructions of the liver, spleen, kidneys, veins, lungs, and abuse of drugs, and stimulating drinks.

There are six different species of dropsy, viz. :—
First.—*Anasarca*, or Cellular Dropsy.
Second.—*Ascites*, or Abdominal Dropsy.
Third.—*Hydrothorax*, or Dropsy of the Chest.
Fourth.—*Hydrocephalus*, or Dropsy of the Brain.
Fifth.—*Ovarian Dropsy.*
Sixth.—*Hydrocele*, or Dropsy of the Testicle.

The accumulation of fluid within the membrane of the spermatic cord is termed Spermatocele, and is of the same nature as Hydrocele.

Treatment.—In treating the different species of dropsy, we must first ascertain the cause, or causes, with a view to their removal. There are two noted conditions to be considered. First, an inflammatory, congestive, or exalted action of the capillary extremities of the arterial vessels, which needs to be treated as all other positive or inflammatory conditions; and, second, the torpor or inactivity of the venous absorbents, which must be restored by general or local tonic treatment.

First, for Anasarca, or Dropsy of the Limbs and Feet.—Place the metallic girdle around the body, over the liver, spleen, and kidneys, to which attach the positive S. C., and, with the negative, treat from the spine down to the feet for ten minutes; then attach the negative to the band, and make passes from the feet upward to the band for

ten minutes more. Next take off the band, seat the patient on positive, and give a general treatment for ten minutes over the whole body, finishing on the kidneys, liver, stomach, pancreas, spleen, and bladder.

Second, for Ascites, or Dropsy of the Abdomen.—Seat the patient on the negative S. C., and, with the positive, treat freely, with a moderate current, over all the parts affected, for fifteen minutes, avoiding much pressure over tender places. Next seat the patient on the positive, and give a general tonic treatment, paying special attention to all the digesting and secreting organs, and finish with positive on the kidneys, and negative over the bladder. Treat daily in this species of dropsy. We must do all we can to promote natural action of the skin, and a healthy secretion in other organs, with normal action of the bowels.

Hydrothorax, or Dropsy of the Chest.—Apply the large chest-plate to the chest, with the positive attached, P. C., and negative on the spine, a little below the positive. Move the plate from side to side, with the current felt slightly; pass the negative over the kidneys, and gently lower both on the spine and abdomen; finish by seating on the negative, and treat with positive all the length of the spine. If the dropsy is about the heart, great care must be taken not to disturb its action: the current must be very gentle, and the treatment only continued for a few minutes at a time, but

may be repeated as often as the case requires. The same rules apply to the action of the skin, and organs of secretion, in this, as the foregoing dropsical affections.

Hydrocephalus, or Dropsy of the Brain.—If treated early, it may be cured. Moisten the hair all over the head; apply a moist cloth, so as to fit closely all over; then apply the metallic cap, with positive P. C., *very gentle*, and treat with the negative all the length of the spine, resting every few minutes, and watching closely the effects. If the symptoms are aggravated, the case is hopeless; but, if relief is experienced, you may expect a cure. Do all in your power to make your patient comfortable, especially keep the room quiet, and well ventilated, and darkened if necessary. Treat for five to ten minutes at a time, and as often as the case requires. Try to preserve a normal action of the skin, and all the organs of the system.

Ovarian Dropsy, Tumor, and Inflammation.—In this species of dropsy, the effusion takes place from the internal face of the membrane which incloses the ovarium, and appears in the form of a small elastic tumor. If there is deranged menstruation, first make that function regular by appropriate treatment. Next seat the patient on the negative P. C., and treat with the positive gentle current over the tumor, and generally over the abdomen, liver, and kidneys; have the patient

avoid all excitement. If there is heat or inflammation in the part, apply cooling moist compresses of arnica, frequently changed. Promote action of the skin and kidneys, and keep the extremities warm, avoiding all exposures. Treat in the same manner for inflammation of the ovaries. If the tumor should continue to enlarge, after treatment for several weeks, the cure may be considered doubtful, and a surgical operation, or tapping, may be necessary as a last resort. The first is dangerous, and the second only palliative. Give treatment once a day, or oftener, as the case may require, and occasionally place the feet in warm water, with the negative.

Hydrocele, or Dropsy of the Testicle.—Treat the enlarged organ by placing it in a cup of tepid water, with the positive S. C., applying the negative to the spine, spermatic cord, and over the kidneys. Treat daily from ten to fifteen minutes, and apply a suspensory bandage if necessary. Avoid lifting and violent exercise, and frequently assume a recumbent position. Bathe the organ with alum-water three times daily, and apply a compress of the same every night on retiring.

For Spermatocele.—Seat the patient on the negative, and apply the positive S. C. gently over the enlargement, for ten minutes, once or twice daily.

HEMORRHAGIA, OR FLOW OF BLOOD.

Hemorrhage may be active or passive.

Active Hemorrhage is generally a result of congestion or increased circulation.

Passive Hemorrhage occurs from a want of vitality, and a relaxed state of the capillaries.

Treatment.—To arrest active hemorrhage, treat with the positive S. C., on or as near the bleeding vessels as possible, applying the negative at some distant or convenient point, and *that* to which you wish to have the blood attracted to, finishing the treatment by a general application, to restore the equilibrium in the whole circulation.

To arrest Passive Hemorrhage, treat as above, until the bleeding ceases; then give general tonic treatment, by seating the patient on the positive, with a view to restore strength and tone to the weak parts, which will need treatment more than the rest, using the negative for that purpose.

CANCER.

Cancer appears under two forms; viz., scirrhous, or occult cancer, and ulcerated, or open cancer. The first is a hard, indolent tumor; the second, a malignant ulcer, discharging a sordid, sanious, or fetid matter.

Treatment.—When we find in the patient a scirrhous diathesis, or that either parent has had cancer, a cure is doubtful; and if the tumor is

indolent, and has not changed much for years, it had better be let alone. But if it is active, and likely to suppurate, or if already an ulcer, then we must do all we can to effect a cure; and our treatment will be both local and general.

In giving local treatment for the tumor, apply the positive P. C. over the tumor, and negative on the spine, a little below the tumor, having the current so mild as not to occasion pain. Treat from five to ten minutes at a time; then seat the patient on the positive, and give general treatment with the negative for ten minutes more. If the tumor becomes sore or inflamed, suspend the first part of the treatment, but continue the second until it passes off. If the tumor has lessened, you may reasonably hope for a cure, and you may treat as before; but do not allow the tumor to suppurate if *possible to prevent it*.

When it is a malignant ulcer, cover it with a moist cloth, and apply the positive on one side, P. C., and negative on the other, keeping the negative nearest the ulcerated part. Treat thus for ten minutes; then apply the negative on the spine, and circulate the positive all around the ulcer for five minutes. Then seat the patient on the positive, and give general treatment with the negative, except on the ulcer, for ten minutes more.

In addition to the foregoing, the patient ought to have a general galvanic bath twice a week, with positive at the feet, or base of the spine, and

negative all over the body, *except on the ulcer*. This is with a view to eliminate from the system all morbid elements, and thus remove the cause of the disease.

In treating cancer of the *womb*, or *vagina*, we use the electrodes for internal treatment on the foregoing principle. For cancer in the stomach, we apply the positive P. C. on the tongue and pneumogastric nerve, and negative on the stomach and liver; current moderate. In all cases, over-exertion must be studiously avoided, and the habits be regular.

TUMORS, GLANDULAR ENLARGEMENTS, AND SWELLINGS.

Tumors are either malignant or non-malignant, encysted or otherwise, and are hard or soft: they are of the consistency of honey, suet, fat, marrow, or flesh, and may result from a scrofulous diathesis, or contusion, or local irritation. If caused by scrofula, a general galvanic treatment should accompany the local, for the purpose of throwing off the morbid elements.

For treatment of tumors, apply the positive P. C. over the tumor, and negative at some convenient point, according to the location of the tumor: in addition, seat the patient on the positive, and give general treatment, both with a view to promote absorption, and favor the natural flow of the currents from the centre of the periphery. Sometimes a metallic plate, of proper shape and size,

makes a good electrode for the tumor. If irritation occurs, discontinue the treatment of the tumor until it subsides, and never have the treatment painful to the patient. Treat as long and as often as the case seems to require. The patient must avoid all excitement, and be regular in all his habits. This treatment applies to swellings of all kinds, and enlarged glands. The general galvanic bath is good in the treatment of tumors, with negative all over the body.

SCROFULA.

Scrofula depends upon a peculiar, depraved condition of the solids and fluids of the system, and gives rise to many diseased manifestations. If hereditary, it is generally incurable, except sometimes in young persons; but, in the worst cases, much improvement attends a thorough course of galvanic and electro-magnetic treatment.

Treatment.—For a scrofulous diathesis, and for all skin diseases resulting therefrom, a thorough galvanic treatment, both in the bath and locally, is the only remedy. In the bath, place the positive at the base of the spine, and treat with the negative, all over the body and limbs, from ten to thirty minutes; then rub the patient dry with a crash towel, if the surface will admit; but, if there it skin disease, dry off as best your can, keeping the patient warm, and in the house for half an hour or more. For local treatment, apply the negative to

the parts affected, and positive at a convenient point, being careful to avoid running the current through the heart, lungs, or any vital organs. The length and frequency of the baths must depend on the patient's re-active power.

Many local affections of a scrofulous nature can be treated successfully by the galvanic or primary current of the electro-magnetic instrument, marked A B, A C, and A D, on Dr. J. Kidder's machine; but A B is more purely galvanic than the rest, and is preferable for treating all kinds of scrofulous sores and glandular enlargements, and those resulting from secondary syphilis, fever sores, lumbar abscesses, hip-joint disease, white swelling (and, for scrofulous opthalmia, positive to the eye, with eye-glass), scorbutus, or scurvy, goitre, rickets, carbuncles, boils, ulcers, and all malignant and offensive skin diseases. All the above class of diseases require to be treated on the same general principles. When there is an inflamed condition, the part requires to be treated with the positive, and generally the negative, on some part of the spine, in the hands, or at the feet; but, if the part to be treated is in the opposite condition, the currents are to be reversed, or the part treated from side to side, or through and through; or the treatment may be varied to suit the requirements of the case. The observing operator can judge better than any written instructions can direct him how to treat all such affections.

EPILEPSY, OR FALLING SICKNESS.

Diagnosis.—The attack often comes on without any premonitory symptoms; but sometimes certain symptoms precede the attack, such as giddiness, dimness of sight, vertigo, starting during sleep, difficult articulation. Some feel a peculiar sensation, called "*aura epileptica,*" like air or water running from the feet and legs, ascending until it reaches the head, when the convulsion sets in; followed by many distorted manifestations, lasting from a few seconds to three-quarters of an hour.

Causes.—The chief are ossification of the brain; depressed portions of bones of the skull, or a morbid condition of the cerebral mass itself; deranged states of the stomach and bowels; repelled cutaneous eruptions; suppressed habitual discharges; debility, and, in females, womb displacement, retained or suppressed menstruations, grief, terror, or fear, and secret indulgence by youth of both sexes.

Treatment.—First ascertain and remove the cause or causes; but, whatever may be the cause, during the paroxysm there is a congested state of the brain. To relieve this, seat the patient on the negative, or, with the feet in warm water, place it there, S. C., and treat with the positive the whole length of the spine, from the neck down to the base, for five minutes; then moisten the hair, and

change to P. C., treating the whole brain from front to back, increasing the current; then pass down the spine again, till consciousness returns; then make the treatment more general, with a view to restore the equilibrium of the whole system. On the next day, make a careful diagnosis of the system, and, in connection with general tonic treatment, give such local treatment as the case requires either to relieve engorgements, subdue local irritation, or restore the normal tone to any part or organ, which must be done according to the principles laid down for the treatment of the different organs and parts of the system; bearing in mind that a perfect equilibrium of all the parts constitutes health, and its opposite disease or *disorder*.

Keep the patient quiet until fully restored. Give treatment once a day, for a week or two, having him avoid all mental excitement and physical exhaustion, and, if a female, all exposures and extremes of temperature. Let the diet be nourishing, and of easy digestion.

APOPLEXY.

Diagnosis.—Very much the same as epilepsy.
Causes.—Very similar.
Treatment.—The same as for epilepsy during the convulsion; afterward give general tonic treatment, and such local treatment as the case seems to require. After having found out the cause, spare no pains to promote its removal, whether it is in

local disturbance or in the improper habits of the patient. Frequently more depends on the patient than the doctor, as much self-denial is often necessary on his part. Restore the equilibrium, and maintain it, and there need be no fears of a second attack.

PARALYSIS, OR PALSY.

Diagnosis.—Paralysis is characterized by a partial or total loss of voluntary motion, or of sensation, or both; it often follows apoplexy, or may arise from disease of the spinal marrow. *Hemiplegia* usually succeeds apoplexy, and *paraplegia* disease of the spinal marrow. *Local palsy* affects some particular part of the body, as an arm, wrist, or face, and may be caused by mechanical injuries, pressure from a clot or effusion, chronic inflammation, or disease of the nervous tissue, or both. Wasting follows loss of voluntary motion.

The perfectly flaccid condition of the muscles of the paralyzed limbs denotes cerebral lesion, distinctly atrophic in its nature, and consequently a negative state of the part of the brain affected; the opposite of inflammatory, in which the vital powers are *below par*. The opposite state of the muscles shows, according to their resistive or rigid state, the degree of irritation or inflammation and lesion of the brain, and evince a plus or positive condition. Paralysis of motion, and paralysis of innervation, occurs generally, not from defect

in the muscular condition itself, but because of an interruption of the passage of the *nervo-vital fluid*.

LOCAL PARALYSIS.

Treatment.—In cases where the muscles are rigid, moisten the hair with cold water, and apply the positive P. C. to the head and neck, with the negative on the spine and over the affected muscles, even of the face. Finish by using S. C. moderate power, using the positive a little above the negative, and treating from the spine to the extremity of the parts affected. But, when the muscles are relaxed and flaccid, treat the brain with the negative P. C. for a few minutes; then change to S. C., and treat as above, with the addition of passing the positive over the muscles and nerves of the well and healthy parts, and negative over the corresponding parts of the other.

HEMIPLEGIA, OR PARALYSIS OF ONE SIDE.

Hemiplegia.—Apply the positive S. C. to the brain, especially the cerebellum; negative on the spine and down to the foot for a few minutes, moderate current; then increase the current, and treat with the positive over the well side to the spine, with negative on the paralyzed parts down to the abdomen. Next pass the positive from the neck down the spine, and treat the whole limb with the negative, and finish by treating the limb through

and through, keeping the positive a little above the negative all the way down to the end of the toes.

Paraplegia, or Palsy of the lower half of the body.—Commence on the head as above, treating all the way down the spine with the positive, having the negative at the feet in warm water, S. C., and finish as in hemiplegia. Treat once a day for a week or two; then three or four times a week. Put a little salt in the water, *after leaving the head*, as it adds to the intensity of the current. Encourage exercise of the will-power over the paralyzed parts; administer frequent and vigorous rubbing and manipulation with the hands, dipped in warm salt water; and let the exercise be moderate, but not fatiguing. Chronic cases often require lengthy treatment,—from one to six months.

CHOREA, OR ST. VITUS' DANCE.

Diagnosis.—Dyspepsia, nervous prostration, constipation, vertigo, palpitation, fulness in the head, confusion, with variable states of mind, spasmodic muscular contraction, growing stronger and stronger.

Causes.—Mental emotions, disappointed love, religious enthusiasm, repelled eruptions, suppressed menstruation, pregnancy, parturition: it often occurs about the age of puberty, and sometimes from vegetable and mineral poisons.

Treatment. — When chorea occurs in young females about the age of puberty, first place the patient in a shallow warm bath with negative at the vagina, and treat with positive S. C., over the lumbar region, ovaries, and pubes, with brisk current for ten or fifteen minutes; then put the feet in warm water, and treat down the spine, and over the same parts down to the middle of the thighs, for ten minutes, at the same time making a few passes over the breasts. Give general tonic treatment every alternate day, for a week or ten days; then two tonic and two local treatments a week until well, or the catamenia is established; in the mean time keeping the patient warm and avoiding all exposures to cold, &c.

When from repelled eruptions, seat the patient on the positive S. C., and treat the whole upper part of the body with the negative for ten to fifteen minutes. An occasional galvanic treatment in the full bath is very good to stimulate excretion, and quiet the nerves. For all other symptoms, give general tonic treatment, and such local treatment as the case requires. If there is mental disquietude, finish by making passes with the hands from the head down the spine. Sometimes a few treatments are sufficient to restore the equilibrium, and cure the patient; at other times, several weeks are required to make a perfect cure; but, by proper electrical treatment, nearly all cases can be cured.

TETANUS. TRISMUS, OR LOCK JAW.

Diagnosis.—Spasm, with rigidity of the muscles of the entire body. The approach of tetanus is generally gradual, and preceded by an uneasy sensation in the chest; spasmodic twitching of the muscles of the throat; difficulty in swallowing : the muscles of the jaws become rigid, until they are immovably locked.

Causes.—The causes of tetanus are numerous, the chief of which are mechanical injuries.

Treatment.—First ascertain the cause, and, if possible, remove it. If caused by a wound, clean it perfectly, and remove any foreign substance that may be causing irritation. Keep the wound open until the contraction is entirely subdued. Treat the wound on general principles; and, to cure the muscular disturbance, seat the patient on the positive S. C., and treat with the negative up the spine and over the chest, neck, jaws, and back of the head, also the shoulders and arms; and if the limbs are rigid, treat from the hips down to the feet, rubbing each part dry as you treat it. Treat every two hours until the muscles relax, and then give tonic treatment until the general health is well established. Keep constantly applied to the wound pledgets of lint, wet with diluted tincture of arnica.

CATALEPSY.

Diagnosis.—Catalepsy; a sudden suspension of motion and sensation. It comes on suddenly, often preceded by languor, vertigo, cephalalgia, flatulent pains in the bowels, depressed spirits, and obtuseness of intellect. When the attack occurs, the body, limbs, and features retain the same attitude and expression as at the moment of seizure.

Causes.—Uterine irritation, especially in young women; irritation of the stomach and bowels; suppressed habitual discharges; close application to study; and violent mental emotions.

Treatment.—When caused by uterine irritation, or obstruction of the menses, treat accordingly, with a view to regulate that function; and so for other local affections: in other words, remove the causes, and catalepsy will cease.

To relieve the attack, seat the patient on the positive, and treat freely, with the negative S. C., from the base of the spine to the cerebellum, and all over the chest, shoulders, arms, abdomen, and lower limbs; continue until the system is restored to its normal state, and give tonic treatment daily for a week or more. For local difficulties, treat until all the functions become healthy and natural.

SYNCOPE, OR FAINTING.

Diagnosis.—Fainting is generally an evidence of a negative condition of the system, when not caused by an organic disease of the heart.

Treatment.—General tonic treatment is what is needed, except when it is clearly traceable to some local cause; and then we treat accordingly, and give tonic treatment besides if necessary. Let the patient be in a reclining position while being treated, if fainting exists at the time of treatment, and bathe the forehead, temples, and face with cold water, also holding a little ammonia near the nose.

VERTIGO, OR GIDDINESS.

Treatment. — Seat the patient on the negative P. C.; moisten the hair, and treat with the positive from the head down over the chest and stomach, finishing with positive on the pneumogastric nerve, and negative on stomach, liver, and bowels. If caused by plethora, pay particular attention to diet; if from suppressed eruptions or dried-up ulcers, give general tonic treatment, with a view to enable the system to throw off the morbid elements.

PALPITATIO CORDIS, OR PALPITATION OF THE HEART, AND HYPERTROPHY, OR ENLARGEMENT OF THE HEART.

Diagnosis. — Sometimes palpitation indicates organic disease, as in the aged, contraction of the

organ, or ossification of its valves. Those of delicate constitution, lax fibre, and nervous excitability, are very subject to it; also the young, from rapid growth.

Causes. — Deranged state of the stomach and bowels; extreme fatigue; over exertion; change of life; pregnancy; any violent mental emotion; too free use of intoxicating drinks; inequilibrium of the circulating fluids.

Treatment. — If we find that the disease results from any local cause, we must treat according to the positive or negative condition of the organ or part affected. If there is enlargement of the organ, ossification of its valves, obstruction, or inflammation, treat with a gentle P. C. positive over the region of the heart, and negative on the spine, a little below the positive. Treat from five to eight minutes, and finish with general tonic treatment, avoiding the heart, and using S. C., with patient seated on the positive. If there is *contraction* of the heart, treat over the heart, with gentle P. C., negative over the organ, and positive over the cardiac plexus on the spine. If from general nervous prostration, general tonic treatment is all that is needed; being careful, in all heart and nervous affections, to begin with a light current, and gradually increase it, according to the condition and strength of the patient. Neuralgia and rheumatism are to be treated as enlargement and palpitation, *but not as contraction.*

ASTHMA.

Diagnosis. — Asthma is of two kinds, — dry and humid. This disease often comes on very suddenly, generally about midnight: the breathing is very laborious, and wheezing; the chest heaves violently; there is an urgent desire for free, fresh cold air; the countenance is expressive of great anxiety; face bloated, livid, or pale; the veins of the head and neck turgid; heart palpitates violently; pulse is irregular and intermitting, and the patient is obliged to keep an upright position; there is cough and expectoration.

Causes. — Asthma is often hereditary, and sometimes depends on a particular constitution of the lungs; violent exercise; cold; changeable atmosphere; sleeping in damp apartments; getting wet when heated by exercise: it also sometimes follows measles, scarlet fever, and other eruptive diseases, and non-appearance of the menses.

Treatment. — If the patient be an adult, and the disease is hereditary, we cannot reasonably expect to produce a permanent cure; but, if the patient is young, there is ground of hope that a permanent cure may be effected. By learning the cause, and removing it, the disease can soon be cured. There is often much contraction of the diaphragm, and the respiratory muscles generally, and partial paralysis of the nerves. Commence by moistening the hair over the cerebellum; treat with the positive S. C. over the cerebellum, pneumogastric

nerve, and down the spine to the cardiac plexus, at the same time using the negative freely over the pectoral muscles and chest, diaphragm, and abdomen; commencing with a moderate current, and increasing it as the patient can bear. Treat for three to five minutes at a time, and rest a moment. This may be continued for twenty to thirty minutes, and, if the paroxysm is on, until entire relief is obtained. It is well, sometimes, to finish with positive on the tongue, and negative on the stomach. Treat daily, or oftener, until a cure is effected, and, if necessary, give general tonic treatment besides: it is good in any case. If asthma is complicated with catarrh or dyspepsia, give special treatment for these diseases. By judicious treatment and perseverance, nearly all cases can be cured.

HEAMATURIA, OR VOIDING BLOOD BY URINE, AND RETENTION OF URINE.

When it passes off without a desire to urinate, it comes from the urethra; when the urine is mixed with blood, attended with pain about the neck of the bladder or perineum, it proceeds from the bladder; but, when it comes from the kidneys, the pain is in the region of those organs.

Causes. — Mechanical injuries; calculous concretions in the kidneys and bladder, or from their lodgment in the urethra; also from cantharides, turpentine, &c.

Treatment.— When the trouble is in the kidneys, seat the patient on the negative S. C., and treat with the positive over the parts affected, and along the course of the ureters to the bladder. If about the neck of the bladder and perineum, treat these parts with positive, and negative at the feet; but, if the trouble is in the urethra, place the genitals in a cup with the positive, and the negative at the feet. Treat from five to fifteen minutes, once or twice a day, till cured. A cooling diet and light exercise will aid the cure : if a recumbent position is most agreeable, enjoy it.

For Retention of Urine. — Place the patient in a shallow, warm sitz-bath, with the negative. Apply the positive on the spine and lumbar vertebra; treat ten to fifteen minutes.

HEMORRHOIDS, OR PILES.

Piles are of two kinds; viz., bleeding and blind piles. Piles are excrescences or tumors arising from the interior portion of the rectum, or situated on the verge of the anus, and are generally very sensitive and painful.

Causes.—Sometimes hereditary; prolapsus ani during childhood, drastic purgatives, obstinate constipation, improper food, excessive drinking, sedentary habits, pregnancy, and parturition.

Treatment.—If there is constipation, first relieve the bowels by tepid injections of castile-soap suds. If the rectum is prolapsed, replace it; then with

the rectum instrument *well oiled*, and connected with the positive P. C., introduce into the rectum about two inches, while the negative is being applied up the spine, and over the stomach, liver, spleen, and bowels. If the instrument cannot be conveniently introduced on account of soreness or swelling, seat the patient on a moist sponge applied to the rectum, and treat as before. In old chronic cases, where the tumors are very hard, a *cautious* use of caustic will hasten their removal. A moist compress, kept on the abdomen during the night, helps to relieve constipation. Particular attention to diet is necessary.

PROLAPSUS ANI.

This is caused by a partial paralysis, and relaxed condition of the sphincter muscle of the rectum, and is a very annoying weakness, but is readily cured by electrical treatment.

Treatment.—First replace the organ; then introduce the rectum instrument, positive S. C., and treat with the negative up the spine, and over the descending colon, on the left side of the abdomen, up to the spleen. In recent cases, a few treatments are sufficient to restore tone to the muscles involved; but, in chronic cases, one to four weeks are needed. If there is constipation, cold injections of castile-soap suds may be used to clear the rectum, and the patient should lie down for an hour or two after treatment. Bed-time is a good

time for treatment, as it favors the natural contraction of the negative muscles.

HERNIA, OR RUPTURE.

Hernia is a protrusion of some of the abdominal viscera from their proper place into a sac formed of a portion of the peritoneum. When it appears in the groin, or labia pudendi, it is called inguinal hernia; when the parts descend into the scrotum, it is called scrotal hernia; when it occurs below poupart's ligament, it is called femoral hernia; when it protrudes at the navel, it is called umbilical hernia; when it appears at any other part of the abdomen, it is called ventral. It is congenital when it has existed from birth. It may be *reducible, irreducible, incarcerated,* or *strangulated hernia.*

Treatment.—First try to replace the parts by careful manipulations; and, if you succeed, then treat over the rupture with the positive S. C., while the negative is on the spine above; this is for the purpose of producing contraction, and restoring tone to the parts relaxed and ruptured. If, by manipulation, you do not reduce the rupture, treat over and above with the positive, keeping the negative on the spine. After the parts are restored, let the patient lie down for several hours; and if, after several treatments, the parts are still disposed to return, apply a suitable truss to keep them in place, until the *contraction* and *heal-*

ing is complete. It may then be carefully dispensed with; but all hard lifting and violent exercise must be avoided. If the parts cannot be restored by the means already tried, a surgical operation may be necessary. If there is heat or inflammation, apply cooling compresses, frequently changed, and moistened by diluted tincture of arnica.

RICKETS.

Treatment.—In treating this disease, we have a double object in view; first to remove the causes which have brought on the disease, and second, to prevent further distortion, and effect a cure.

Sometimes the first is impossible, as the cause is hereditary; sometimes the second is possible; in either case, we depend chiefly on general tonic treatment, and such local treatment as the case seems to require, whether it be curvature of the spine, protruding sternum, or enlarged joints. Nutritious food, strict cleanliness, fresh air, and moderate exercise, and frequent hand-rubbing from head to feet by a healthy magnetic person, are very necessary to aid in the case. Treat locally on general principles, as the condition and symptoms require.

ERYSIPELAS, OR ST. ANTHONY'S FIRE.

This is a febrile disease, attended with inflammation of the skin when it appears externally, and of the mucous membrane when it is internal.

The parts attached are in a highly positive condition: hence the treatment must correspond.

Treatment.—When the parts affected are above the hips, seat the patient on the negative, and treat the parts affected with the positive S. C. If on the head, moisten the hair, begin with a gentle current, and increase it cautiously: don't have it feel painful. Treat from five to fifteen minutes, and repeat every few hours until the inflammation subsides. If it is internal, place the positive on the tongue, with the negative as above. If it is in the limbs, put the negative at the feet, and treat with the positive over the affected parts with moderate current.

Powdered flour on the surface, or a covering of silk oil-cloth, is sometimes very desirable. No stimulating food and drink should be taken: diet must be plain. Care must be taken to avoid cold and damp atmosphere. If taken early, sometimes one or two treatments will arrest the disease.

MENTAL DERANGEMENTS.

This condition has many phases of manifestation, and many causes producing it. Our object is first to ascertain the causes, and then as far as possible remove them; then restore the mind, and any organs or parts involved, to their normal condition. Sometimes the treatment is almost entirely mental, and consists in kind persuasion, and friendly but positive assurances, that, by pursuing

a certain course, perfect restoration will take place. But, when the mental powers are so deranged that reason cannot exercise its true functions, we must do the best we can otherwise.

Treatment.—When there is organic derangement, we must treat the organ or organs affected according to general principles. On a careful examination of the head, we shall probably find one part much hotter than the rest, which indicates too much action in that part, while at the same time the extremities may be cold: this shows an unbalanced circulation. To restore equilibrium, we must moisten the hair, apply the positive P. C., to the head, and place the negative at the feet, and, in addition, treat with the positive all down the spine. If there is soreness on any part of the head, treat with a very *light* current, and do not hold the electrode steadily on those parts. Treat the head all over, and increase the current when treating down the spine, but do not have it painful anywhere.

If there is emaciation and general nervous prostration, alternate the local with general tonic treatment. Treat once a day, or oftener, as the case seems to require. Moderate exercise in the open air, cheerful company, pleasant social surroundings, with objects to prevent the mind dwelling on itself, are among the conditions necessary to a restoration. Wisdom and tact on the part of the physician are very necessary to suc-

cessful treatment in these cases, as no two will require precisely the same treatment. Imbecility in youth and children can often be cured by judicious and persevering treatment, especially before the faculties of the mind have attained their growth.

MANIA-A-POTU, OR DELIRIUM TREMENS.

The causes and symptoms of this disorder are generally pretty well understood. There is great nervous prostration, with mental and gastric derangement.

Treatment.—Seat the patient on the negative, moisten the hair, and treat the head with the positive P. C., for a few minutes at a time for ten minutes; then increase the current, and treat down the spine, and over the stomach, liver, and bowels. Treat twice a day with general tonic treatment at night, omitting to treat the head; but the positive may be applied to the tongue, and the negative to the stomach. Keep the patient quiet as possible; let the diet be nutritious but not high seasoned or stimulating, as the stomach is very weak and sensitive.

HYPOCHONDRIA.

In this disorder, the mind is in direct sympathy with the stomach, liver, and spleen: there is generally great nervous prostration, with gloomy

forebodings in relation to the future, and indisposition to any undertaking.

Treatment.—Commence by giving general tonic treatment, and alternate by giving local treatment to the stomach, with positive S. C. over the pneumogastric nerve, on the upper part of the spine, and negative over the stomach. Then apply the positive along the spine above the stomach, and negative over the liver, spleen, stomach, and bowels. If there is constipation, seat the patient on the negative, and treat with the positive over the ascending, transverse, and descending colon, thus finishing the treatment. If you treat twice a day, give the tonic treatment in the morning, and local in the evening, one hour before eating. Everything cheering and pleasant, with open-air exercise, should be enjoyed by the patient.

HYSTERIA, OR HYSTERICS.

When we have reason to believe the cause to be in some deranged action of the reproductive organs, we must treat according to the nature of the deranged function. But, if we find it to be from nervous prostration, we must give general tonic treatment, and such mental treatment as in our judgment the case requires, and endeavor to establish and preserve a true balance of all the electrical forces. Our success in treating this affection will depend on our knowledge of the temperament of the patients, and adapting the

treatment to their peculiar condition. An occasional galvanic bath will aid much in the cure.

ALOPECIA, OR FALLING OFF OF THE HAIR.

This often occurs after fevers, especially if much mercury has been taken. Protracted grief, severe headaches, debility from loss of animal fluids, and syphilis, are the chief causes of this affection. The blood and other fluids have been deteriorated, and the system is in a negative and impaired condition.

Treatment.—Every one knows how readily the hair is affected by electricity, which of course is from the connection of its roots with the nerves of nutrition. To restore tone to the whole nervous system, and especially to those which supply vitality to the hair, we give general tonic treatment to the whole system, by seating the patient on the positive, and treating with the negative S. C. up the spine, and over all the vital organs, finishing with a light current on the head, after first thoroughly moistening the hair, and if necessary washing the scalp thoroughly with castile-soap suds, to cleanse the skin and open the pores. A treatment once a day is sufficient. An occasional bath, nutritious food at regular intervals, and moderate out-door exercise, will soon restore the hair, and all the other functions of the organism.

DEAFNESS.

As all are aware, the ear is a delicate organ, and is liable to many diseased conditions from numerous causes. Among the first which claim our attention are inflammations, external or internal, acute or chronic, and all painful conditions, from whatever cause. These have all to be treated on the same general principle: as all painful conditions show that there is more or less irritation, or inflammation, and a too *positive* condition of the parts affected.

Treatment of inflamed or painful conditions.— First ascertain the cause, if possible; and if it is from the presence of any foreign body, hardened ear-wax, or excrescences, examine the ear by means of an ear speculum, and with a pair of forceps *carefully* remove the foreign body; or with an ear-spoon remove as much of the hardened wax as possible. Then syringe the ear with tepid water till the wax is all removed. Excrescences will have to be removed either by caustic, ligature, or the knife. To remove pain, and subdue inflammation, apply the positive P. C. to and around the ear, by means of a sponge electrode, and internally by means of the ear instrument, tipped with soft, moist sponge; or the cavity may be filled with tepid water, and the tip of the ear instrument just inserted in the water. This distributes the current all through the nerves, both of the meatus, and internal ear. While

treating thus with the positive, let the negative be placed on the back of the head towards the opposite ear, on the tongue, and in front of the root of the nose, and finish with the negative on the spine, below the seventh cervicle vertebra. This treatment should be continued daily, until all pain and soreness is removed. Afterwards, if there is any discharge from the ears, treat in the same manner, but with the negative where you applied the positive before. Begin with a very light current, and always avoid having it painful. Treat from five to fifteen minutes, and protect the ears by a little fine cotton wool on going into the cold air.

Deafness, from paralysis of the acoustic nerve, is treated in the same manner as for a discharge from the ear, with the addition of general tonic treatment when there is debility and general nervous prostration.

CEPHALAGIA, OR HEADACHE.

Nervous Headache.—Treatment.—If caused from a rush of blood to the head, moisten the hair, and treat over the head with positive P. C., by means of the metallic cap, moist hand, or sponge electrode, at the same time treating with the negative over the stomach, liver, spleen, and kidneys, for a few minutes; then bring the positive on the spine, and increase the current, and treat for ten minutes longer, passing the positive down opposite the

stomach. If there is constipation, next seat the patient on the negative and treat freely, with a little pressure all over the bowels, following their course from the stomach to the descending colon, on the left side, as low as the pubes. Finish by manipulating the head with the hands, without the current, which will equalize the nerve forces, and soothe the brain.

Sick Headache is generally caused by derangement of the stomach and liver, sometimes called bilious.

Treatment.—Apply the positive to the pneumogastric nerve, just below the cerebellum, S. C., and treat over the liver, stomach, and spleen with negative, occasionally passing the positive down the spine. If the pain does not cease, moisten the hair, and give the head a light treatment with the moist hand, positive, and P. C., and finish with manipulation by both hands, making the passes downward along the spine.

If there is constipation, treat as in nervous headache. As there are many causes of headache, and some of them peculiar to females, the foregoing treatment may be modified to meet the case: keep the extremities warm by appropriate treatment.

NOISES IN THE HEAD, RINGING IN THE EARS, ETC.

This may be caused by nervous prostration, congestion of the capillaries, or the blood being too thick, and not circulating freely.

Treatment.—If from nervous prostration, give general tonic treatment, with local treatment to the head, as in case of deafness. If from congestion, moisten the hair, and treat the head and ears with the positive, and negative on the spine. If the blood is too thick, give tonic treatment to the liver, kidneys, spleen, and surface generally. Drink freely of pure, soft water; bathe often, and take exercise freely out-doors, and not think of the disease any more than can be avoided.

SPINAL CURVATURE.

Spinal distortions often result from organic affections, caries or injuries of the vertebral column, or from osseous malformation, as in rickets and scrofula; but most are the result of muscular debility. In no class of diseases has medical practice proved itself more inadequate to cure than in this. All through the civilized world, unfortunate cases abound, especially females, whose backs have been blistered, burned, scarred, cauterized, leeched, cupped, scarified, and otherwise treated, with the view of counter-irritating a spinal disease, when the only trouble was nervous prostration, or muscular debility. In this condition, the vertebral column cannot be sustained erect; so it bends, leans, or tips backward, forward, or to one side,—generally the latter,—and no artificial supports or drugs will remedy the evil. *Tone* to the nerves and muscles, locally and generally, is

what is needed. "The small of the back" is the centre of the whole muscular system; and no less than three hundred distinct muscles are concerned in the complicated movements of the vertebral column. Thence we can easily understand how a relaxed or weakly condition of the general system should contribute to a muscular distortion of the spine.

In organic or structural derangements, the distortion is from within outward, forming a sharp projection of the bones, called "angular curvature." This often causes paralysis of the lower extremities, and is seldom curable; whilst the other, or lateral curvature, can always be cured if attended to in season: I have cured such of twenty years' standing.

Treatment.—Do everything that will improve the general health, and invigorate the system. Give general tonic treatment, especially if the system is debilitated, for several days, or a week; and, if the curvature is on the upper part of the spine, since our object is to cause contraction of the relaxed muscles, and relaxation of the contracted ones, and as we know the positive electrode causes *contraction*, and the negative *relaxation*, by means of two insulated sponge-handles, we treat over the contracted muscles with the *negative*, and over the relaxed ones with the *positive*, thus restoring the lost balance between them, and curing the curvature. It is well to apply a brisk

current, and apply good pressure, especially on the convex side, thus aiding nature in adjusting the cartilages and bones; for, in chronic cases, the cartilages are wedge-shaped, and often require a little artificial aid till they become adjusted. The treatment may be varied by passing one electrode down the whole length of the spine, while the other is used on the muscles that are contracted; and then change, by treating the spine with the other, the relaxed muscles being treated as before. Sometimes the curvature is double, and shaped like the letter S; treat each part as a single curvature. The patient must be urged to stand, sit, and lie, in such positions as will most favor the natural shape of the spine. A hair mattress makes the best bed; feathers must not be used. Treat daily, for a week or two, fifteen to thirty minutes at a time, using S. C. If there is spinal irritation, relieve that first; but, if there is caries of the bone, a cure is doubtful. Recent cases in young people are soon cured; but old chronic cases take much longer. All cases can be benefitted.

SPINAL IRRITATION.

Treatment.—Seat the patient on the negative S. C., and treat with positive over the sore place, and also over the whole spine. If the irritation is very low on the spine, place the negative between the knees, or at the feet, and graduate the current

according to the feelings of the patient, and vary the treatment as the case requires. Treat daily till relieved.

HEPATIZATION OF THE LUNGS.

This term is applied to the lungs when gorged with effused matter, so that they are no longer pervious to the air, but resemble liver in appearance and consistency. It is a sequence of pneumonia, and is to be treated in the same manner, except that the *positive*, and not the negative, is to be used over the hepatized part, and negative always below the positive.

NEURALGIA OF ALL KINDS.

By the term neuralgia, we designate all those painful affections, in different parts of the body, of a purely nervous character. When in the face, *tic douloureux;* in the stomach, *gastrodynia;* in the back, *lumbago;* in the hip and thigh, *sciatica;* in the feet and legs, *neuralgia pedis,* &c.

Treatment.—When the pains are not local, but appear in different parts of the body, it shows general deranged nervous action and prostration, and requires general tonic treatment, and a removal of the causes as far as possible.

When the pain is local, and attended with sharp, lancinating pains, and redness and fulness of the blood-vessels in the part, treat with the positive

from the nervous terminations to the spine, and with the negative on the spine, moving it slowly from the roots of the affected nerves downward, if in the upper part of the body; and at the coccyx, if in the back; and at the feet, if in the hips and limbs.

But if we find the parts affected to be cold and pale, with a soft, flabby state of the muscles, and tendency to atrophy, or wasting, then we know it is a negative condition, and requires the opposite treatment, alternating with general tonic treatment, and such other measures as will tend to improve the general health. Gentle manipulations over the parts with the warm hands are often of great service; and the more magnetic and healthy the operator, the better. As the healing of the patient is our first and only object, we have no "iron-bedstead rule" by which to do it. When the eyeball is the seat of the disease, we use the eye-glass; and so of the tongue or the ear. Sometimes I have found that hot sponge electrodes had a more soothing effect than cold. But the parts must be protected from exposure to cold immediately after treatment.

DIABETIS.

Diabetis is an immoderate flow of urine. There are two varieties; viz., *diabetis insipidis* and *diabetis mellitus*. In the former, there is an excessive discharge of urea; in the latter, large quantities of

sugar. In *diabetis mellitus,* there is much derangement of the digestive functions, with morbid appetite, skin dry and harsh, weakness in the loins, debility, constipation, extremities cold, swelling of the feet and legs, with some inflammation and pain at the orifice of the urethra.

In *diabetis insipidis,* the thirst, and desire for food, are less, skin not so dry, some pain in the back, with some irritation at the neck of the bladder, extending along the urethra.

Causes.—Exposure to cold ; too free indulgence in highly-seasoned food and spirituous drinks; mercury, and violent mental emotions.

Treatment.—Since the primary cause is in the derangement of the digestive and assimilating functions, we must try to restore a healthy tone to these departments. First, place the positive S. C. on the root of the pneumogastric nerve, and treat with the negative over the stomach, liver, spleen, pancreas, and bowels, for ten or fifteen minutes. Then seat the patient on the negative, and treat with the positive over the kidneys, and along the course of the ureters, down to the bladder. Next seat the patient in a tepid sitz-bath, having the water over the pubic arch, placing the positive in the bath, while the negative is at the feet. This is to reduce the irritation in the bladder and urethra. Finish with the negative along the spine, and positive over the feet and limbs, up to the hips ; that is, if there is swelling in the feet

and limbs; if not, the last part may be dispensed
with. Let the current be as strong as agreeable to
the patient.

The preceding treatment should be alternated by
general tonic treatment, as all diabetic patients are
more or less debilitated. If the patient is too weak
to have all the treatment spoken of administered
at one time, it may be given part at a time. Treat
daily, or oftener if needed, remembering that the
amount of treatment must correspond to the
patient's re-active or recuperative power; and
sometimes this is very limited. Let the diet be
light, nutritious, and of easy digestion, with salt
enough to suit the taste. For drinks, give water,
milk and water, and mucilaginous teas. Take
moderate exercise in the open air, and avoid gloom
and despondency. Be cheerful, hopeful, and
happy as possible : this aids the cure.

ASPHYXIA, OR SUSPENDED ANIMATION.

Asphyxia may be caused by submersion, by a
stroke of lightning, by the inhalation of poisonous
gases, or by a gradual loss of power in the respiratory muscles, and from strangulation.

Treatment.—Our first effort must be to restore
respiration. Apply the positive S. C. to the
pneumogastric nerve, and negative over the chest
to the diaphragm and over the respiratory muscles
of the abdomen. Have the current strong enough
to produce contraction and expansion of the mus-

cles. Make momentary pressure on the chest; then quickly remove the hands, and allow the chest to expand. Continue the treatment in both ways until the breathing is restored, or you conclude the case is hopeless. Have the room well ventilated, and keep the body warm by rubbing, and other artificial means. After restoration, give general tonic treatment.

Asphyxia of New-Born Infants.—Clear the mouth; then gently blow into the mouth and nostrils; press moderately on the chest, to expel the air. Continue till resuscitation is established, or you see there is no ground of hope. A gentle stimulus to the tongue and nose is often of great service.

RECENT WOUNDS, CONTUSIONS, AND BRUISES.

Treatment.—If there is extravasated blood, discoloration, and soreness, we treat with the positive P. C. over the parts, both to produce contraction of the engorged vessels, promote absorption, and subdue irritation. Apply the negative on some proximate healthy part. If the surface of the wound is broken, cover it with a moist cloth, and apply the sponge electrode over that.

BOILS AND FELONS.

In their early stages, these are to be treated on the same principles as wounds, &c., except that the hand or finger may be put in a bowl of water with the positive, negative on the spine, or in the oppo-

site hand, or at the feet. If taken early, suppuration may be prevented. The treatment may be repeated two to six times a day, according to the requirements of the case.

If suppuration takes place, it may be hastened by reversing the treatment. An incision made to the bone is often the most prompt way to cure a felon. Cooling applications may be applied with benefit; and, to promote healing, a compress moistened with diluted tincture of arnica, aids the process. To prevent suppuration, we apply the positive to the swelling; to hasten it, we apply the negative.

CANKER IN THE MOUTH, AND SORE MOUTH.

Canker is a condition resulting from a depraved state of the fluids of the body, and requires general tonic treatment to remove the cause; but, for the mouth, apply the positive P. C. to the tongue, and treat with the negative over the sides of the face, throat, and neck. What is called sore mouth is to be treated as canker in the mouth.

BUNIONS, CORNS, AND CHILBLAINS.

Apply the positive to the enlarged and inflamed part, and negative on some other part of the foot. Pare the corn, and with the ear-handle, tipped with a bit of sponge moistened in water, treat the corn, positive on the corn, negative on some

other part. Remove all the causes, so that the effects may cease.

TO RESTORE FROZEN EARS, LIMBS, ETC.

Seat the patient on the positive P. C., and treat the parts affected with the negative. Continue till the circulation is restored.

CRICK IN THE NECK.

This is a painful rheumatic affection of the muscles of the neck, which causes a person to hold his head on one side in a characteristic manner.

Treatment.—Apply the negative S. C. over the contracted muscles, and the positive on the opposite side. Next pass the positive along the spine, over the dorsal region, with negative as before. We sometimes place the positive on the tongue also; but this only admits of a very light current, and is not so effectual as the other modes.

CRICK IN THE BACK.

This is to be treated by seating the patient on the positive, treating with the negative along the spine over the part affected. This is rheumatic; and to remove the disease, and prevent a recurrence, we need to give general tonic treatment over the stomach, liver, kidneys, and spleen.

CRAMP OF LIMBS, STOMACH, ETC.

Cramp is a local spasmodic contraction of the muscles of a part. Apply the positive above the

part affected, and negative over the contracted muscles. If in the stomach or any vital organ, apply the positive over the roots of the nerves, and negative over the organ. Commence with light current, and increase as required.

CIRCULATION OF THE BLOOD.

Treatment.—To promote a healthy circulation of the blood, especially in the skin and extremities, apply the positive S. C. to the seventh cervical vertebra, and treat with the negative down the spine, over the shoulders, down the arms to the hands, then over the chest, loins, and abdomen; then apply the positive over the lumbar vertebra, and negative down the limbs to the feet. Sometimes it is well to place the feet in a warm footbath, negative in the water. Brisk hand-treatment should follow the electrical, and plenty of out-door exercise, with occasional general bathing and rubbing, which will aid much in the restoration.

GENERAL DEBILITY, NERVOUS AND OTHERWISE.

This condition is induced by long and constant over-taxing of the system. It is an "all-run-down" condition, and requires general tonic treatment. Seat the patient on the positive S. C., and treat with the negative all over the body and limbs, finishing on the cerebellum, and with a light current over the forehead and temples. Bathing and

rubbing, as in the preceding, are good here. Treat daily for a week or two, then three times a week. Patients generally feel better in a few days; but they must continue until restored.

OREANA, OR ULCER OF THE NOSE.

This is a fetid ulcer in the nose, sometimes malignant, accompanied with caries of the bones, and involving the antrum, frontal sinus, and adjacent structures.

Treatment.—Fill a washbowl two-thirds full of tepid water; put in the negative electrode P. C. and a large sponge; let the patient immerse the forehead, and as much of the nose as possible, and not have the water enter the nostrils. Treat with the positive up the spine to the cerebellum for ten to fifteen minutes; then take a small electrode, tipped with moist sponge, and treat inside the nostrils for a few minutes. Finish by treating the root of the nose and forehead over the frontal sinus with moist sponge electrode, and, if necessary, alternate with general tonic treatment. Treat daily if necessary.

POLYPUS.

This is a tumor in a cavity of the body, as in the nostrils, uterus, vagina, &c.

Treatment.—Apply the positive P. C. on or over the tumor if possible: if in the womb or vagina, by

womb or vagina instrument; negative on the spine opposite. If in the nose, positive on, or in the nose; negative on the spine or back of the neck. Treat daily until removed, or you find a surgical operation necessary.

SLEEPLESSNESS.

This condition has many causes, and a little investigation will usually reveal them. Having learned the causes, we must, if possible, remove them, and by soothing and tonic treatment restore the equilibrium. Generally the mind and body have been overtaxed, or prostrated by disease: in any case general tonic treatment is needed. This should be given from the first, and *in the morning*. An hour before retiring, place the feet in warm water, with the negative; moisten the hair, and, with the moist hand as the positive electrode, treat the head with a very light current, just perceptible to the patient, for seven minutes; finish by increasing the current a little, and treat with *positive* down the spine. Tell the patient to feel no anxiety about sleep, for that tends to prevent sleep.

STAMMERING.

When owing to serious malformation of the tongue or other organs of speech, it may be incurable; but the disease is generally spasmodic,

and often easily cured. It is well to fill the chest well before beginning to speak; *avoid repeating syllables or words ;* but, if necessary, utter words or syllables slowly, to a steady beat of the hand.

Treatment.—Apply positive on the tongue, S. C., negative over the organs of speech and all the muscles used by the vocal organs, and over the chest. Next place the positive on the dorsal vertebra, and treat with negative up the spine, and all over the muscles of the neck, jaws, and chest. If there is general debility, give general tonic treatment.

WEAKNESS OF STOMACH, LUNGS, HEART, AND FOR VOMITING AND WATER-BRASH.

Treatment.—As the pneumogastric nerve sends branches to all of these important organs, we treat by applying the positive S. C. to the roots of these branches, and treat with the negative from the spine over each organ; but especial care is needed to begin with a light current, *especially over the heart.* General tonic treatment is always beneficial in these cases. In case of vomiting and water-brash, vary the treatment by placing the positive on the tongue, while the negative is being applied to the stomach, liver, duodenum, and bowels.

SUNSTROKE.

This is attended with many symptoms of congestion of the brain, and requires prompt treatment.

First moisten the hair, and treat with metallic cap and positive P. C., very mild, on the whole head, and down the spine, with negative over the stomach and liver; finishing with negative at the coccyx, and positive down the spine. If the stroke is recent, treat three or four times daily. But, if the effects are of long standing, alternate the foregoing by tonic treatment, treating lightly over the back brain, or cerebellum, for a few moments each time. Avoid all exposure to the hot sun, and preserve as much as possible an equilibrium of all the forces. Be temperate in all things, especially avoiding mental excitement.

TOOTHACHE.

This arises from various causes. When from cold, place the positive on the tongue, as near the tooth as possible, and treat with the negative P. C. over the face and neck and chest. But if it is from caries or ulceration, then apply the positive directly to the tooth, with current strong enough to relieve the pain, continuing to treat from ten to fifteen minutes, and repeat every two hours until cured. If pregnancy is the cause, seat on the negative, and treat with the positive over the face and cerebellum, and down the spine. Cease as soon as relief is obtained. If more convenient, the negative may be held in the hands, instead of being at the coccyx.

URINARY CALCULI; GRAVEL OR STONE IN THE BLADDER.

The term gravel is generally applied to the sand-like concretions, like brick-dust or small stones, which form in the kidneys, and in a few days pass through the ureters to the bladder, causing shooting pains down the back, through the pelvis, and in the thighs, numbness of the legs, and a retraction of either testicle. When the calculous formation has acquired a size which renders it difficult to pass from the kidney or bladder, it is then termed stone, and is known by the following symptoms. Frequent inclination to urinate, with severe pain, voiding it drop by drop. Sometimes the urine will start in a full, natural stream, and suddenly stop, from the calculi rolling against the orifice of the neck of the bladder. The urine is turbid, bloody, or slimy, accompanied with a distressing, bearing-down sensation in the bladder, and sometimes, by sympathy, causing, prolapsus ani.

Causes.—Of the causes of this disorder, medical men are not fully agreed; but among the most certain is a gouty diathesis, hard water, saleratus in food, strong acids, fermented liquors, much salt, high living and gormandizing, and whatever tends to impoverish or deteriorate the blood, and derange the action of the stomach, kidneys, liver, or bowels, or the functions of the skin and lungs.

Treatment.—First correct all the habits of life,

and do all in your power to favor the normal action of all the functions of the organism. To aid in restoring healthy action to all the organs (brain included), give general tonic treatment. To correct the morbid and gravelly secretion from the kidneys, seat the patient on the negative P. C. and treat with the positive over the region of the kidneys, and down the course of the ureters to the bladder, wherever there is soreness to the touch. Treat also over the liver, spleen, stomach, and bowels, at the same time.

During the passage of calculi along the ureters, treat with positive over the kidneys, and negative over the ureters, as this tends to relax the tube and relieve the pain. Have the sponge electrode as moist and hot as it can be endured: hot fomentations are also good.

To disintegrate stone in the bladder, pass into the bladder a metallic catheter, insulated to within an inch of the end. Bring it in contact with the stone, P. C., and negative pole, and with the positive treat over the bladder, on the spine and sides, or on any other part that will enable you to pass the current through the stone. Have the current of such strength as can be endured without much pain. Treat in this way for half an hour; then withdraw the catheter, and place the genitals in a cup of warm water with the negative, and treat with the positive over the kidneys, spine and bladder, for ten minutes. Treat once or twice

daily, and observe if much of the stone is passed in urinating. If there is pain and irritation about the orifice of the urethra, place the penis in a cup, with the positive and negative at the feet; treating thus for five minutes.

In treating this disease, much skill and tact is necessary, and a correct anatomical knowledge of the parts, as an improper handling of the catheter might cause much unnecessary pain, and possibly inflammation of the urethra.

BRIGHT'S DISEASE OF THE KIDNEYS.

As the cause of this disease is in the impurity of the blood, we must remove that by proper attention to diet, bathing, and exercise in the pure air, and all means within our reach to effect that end. If it is from suppressed eruptions, we must do all we can to restore the normal action of the skin, and thus relieve the kidneys; or, whatever are the disturbing causes, try to remove them.

Treatment.—Commence with general tonic treatment, and treat locally with positive P. C. over the kidneys, and negative at the coccyx, following the course of the ureters to the bladder, and, in addition, treat the stomach, liver, and spleen in the same manner. If the disease is attended with dropsical effusions, treat as in dropsy; if fever, treat as for fever; and so as regards any local or general symptoms: for many of the symptoms are like those attending both inflammation of the blad-

der, inflammation of the kidneys, and urinary calculi; this being the case, we must vary the treatment according to the symptoms.

RETENTION OF THE MENSES.

The menses not appearing at the proper period, is generally attended with deranged action, and the general health is soon impaired. The following symptoms are very common: dizziness, faintness, determinations to the head, bleeding at the nose, difficult breathing, palpitation of the heart, peculiar cravings, nervous prostration, sometimes convulsions, and St. Vitus' dance, ending with consumption.

Treatment.—Begin with general tonic treatment. Treat *locally* with negative S. C. under the patient, in a shallow warm sitz-bath, having the positive attached to the metallic girdle around the waist; or treat with the sponge electrode over the dorsal and lumbar region, and also over the abdomen, *on* the ovaries, and over the womb: treat in this way for twenty minutes. Occasionally place the feet in warm water with the negative, and treat with the positive as before. Treat once a day, and alternate the general tonic with the local treatment. I have never known this treatment to fail if persevered in. If the courses do not appear within a month, and the general health is improved, suspend the local treatment for a week or two, then treat again. Keep the feet warm, take out-door

exercise, and let the diet be nutritious and easy of digestion; take a warm foot-bath on going to bed, and occasionally a warm sitz-bath at the same time.

CHLOROSIS, OR GREEN SICKNESS.

As this disorder is occasioned by retention of the menses, the same treatment is appropriate, except when there is palpitation of the heart. In that case, the current must be applied very gently in the region of that organ; and, if it is only sympathetic, general tonic treatment is all that is needed to regulate the heart. Diet, exercise, &c., as in retention.

AMENORRHŒA, OR SUPPRESSION OF THE MENSES.

This implies a "temporary cessation of the menses:" headache, dizziness, and congestion of the head, chest, and stomach, are frequent symptoms.

Causes.—Chiefly cold from wet feet, violent mental emotions, general prostration, and improper diet.

Treatment.—If there is congestion of the head, chest, or stomach, seat the patient in a warm sitz-bath, with the negative S. C., and treat with the positive from the cerebellum, down the spine, over the lungs and stomach; then over the lumbar and sacral nerves; and lastly over the abdomen, ovaries, and uterus. Alternate this treatment by placing the feet in warm water with the negative,

and treat the spine, abdomen, ovaries, and uterus with the current positive, pretty strong. Diet light and cooling.

DYSMENORRHŒA, OR PAINFUL OR DIFFICULT MENSTRUATION.

This disorder is accompanied with severe pains in the back and loins, and also in the ovaries and abdomen. With some, it seems constitutional; but cold, and improper living, is often the cause.

Treatment.—As in amenorrhœa, with the addition of the vaginal instrument applied internally, with negative and positive on the lumbar nerves, and over the ovaries and uterus. This may also be applied with advantage in amenorrhœa when the case requires it.

LEUCORRHŒA, FLUOR ALBUS, OR WHITES.

This affection is very debilitating. It is caused in various ways, but chiefly by uterine displacements, exposures to cold, laxity of parts, mechanical injuries, and uterine diseases.

Treatment.—When leucorrhœa is caused by irritation, apply the vaginal instrument, with positive P. C., internally, and negative along the spine, over the abdomen, hips, ovaries, and uterus, finishing on the spine, between the shoulders. But if it is caused from relaxation and general debility, and of long standing, treat internally, with the nega-

tive and positive on the spine, &c.; and give, besides, general tonic treatment. Considerable time is often required to cure chronic cases. The whole treatment may occupy twenty minutes each day. To aid the cure, use tepid vaginal enemas of castile soap and water two or three times daily.

MENORRHAGIA, OR IMMODERATE FLOW OF THE MENSES, OR FLOODING.

This is flooding, or hemorrhage from the uterus, and characterized by pain in the back, loins, hips, and abdomen. It is caused by violent exertions, falls, active cathartics, dancing, and everything that tends to debilitate.

Treatment.—This disease is somewhat dangerous from liability to excessive loss of blood, and requires prompt and efficient treatment.

Connect the womb-director with the positive S. C., and bring it in contact with the os-tincia, or mouth of the womb; apply the negative on the spine and upper part of the abdomen. Commence with a light current, and gradually increase, until it is sensibly felt in the uterus. Treat for ten minutes in this way, and then wait to see the result. It is not always best to stop the flow suddenly, as it might cause congestion of the brain, or some vital organ. Treat from one to three times daily, and after the flooding ceases give general tonic treatment by seating on the positive. Diet nutritious, and easy of digestion.

IRREGULAR MENSTRUATION.

For premature appearance of the menses, give general tonic treatment, commencing soon after the flow ceases, and continue until next time. As relaxation or general debility are the chief causes, the above treatment is very appropriate; but it may be varied according to the symptoms or requirements of the case.

For delay of the menses, and when they do not continue long enough, treat as in painful menstruation, having recourse to the warm hip and foot baths.

When the discharge continues too long, treat as in menorrhagia, or flooding, giving, also, general tonic treatment; and, if not from plethora, let the diet be nutritious and of easy digestion; and in all cases keep the bowels, kidneys, skin, and other functions of the body, as perfect as possible.

CESSATION OF THE MENSES, OR CHANGE OF LIFE.

This change occurs at about the forty-fifth year, and is often attended with many disturbances of the different functions of the body. It may truly be said to be the "critical period;" but, as it is a *natural change,* if we have lived in harmony with nature's laws before, it is seldom attended with danger. Our treatment is *not to prevent it,* but rather to hasten and smooth the way to its consummation; and, to do this in the best manner, our

treatment will have to be varied according to the condition and symptoms of the patient. As they are so varied, no precise instruction can be given, except to treat according to the general principles of our philosophy and mode of treating diseases and unbalanced conditions.

PREGNANCY, OR UTERO-GESTATION.

Although utero-gestation is a natural process, yet there are many conditions connected with it that may be greatly modified, and much comfort insured by a little judicious treatment; but here, again, the amount and kind will depend on the condition and symptoms. Yet, in all cases of debility and nervous prostration, general tonic treatment is always beneficial. To remove spots on the face, apply the positive P. O. to the spots and face generally, and negative on the kidneys, liver, and spleen, with weak current.

ABORTION, OR MISCARRIAGE.

As all are aware, this is liable to occur at any time from conception to the seventh month; and after that time it is called premature labor. Now, as electro-magnetism is admitted to be the greatest emenagogue in nature, we know by its positive and negative polarities that it is capable of producing the greatest muscular contraction or relaxation of any known agent. Therefore, in threat-

ed abortion, we need its prompt contracting power. Apply the positive S. C., by means of the womb or vaginal director, to the mouth of the womb, or to the vagina and womb by means of the vaginal instrument; at the same time applying the negative to the spine and upper part of the abdomen. In addition, cold compresses may be applied over the womb; for, upon the arrest of the hemorrhage, and firm contraction of the uterine vessels, does the safety of the fœtus depend.

On this principle it will be evident, that by reversing the poles, the opposite results are likely to follow, for abortion has been known to follow its application. Of course the persons most liable are those of lax fibre, nervous temperament, or great debility. Great caution is therefore needed in its application, being sure that the positive and negative are applied to the right parts; for the most dangerous feature in these cases is the hemorrhage. If the miscarriage has gone too far to be arrested, the same treatment may be applied after the fœtus and placenta are discharged, to prevent or arrest hemorrhage. Cold compresses on the abdomen and cold vaginal injections are often of great service to prevent abortion.

PARTURITION, OR DELIVERY.

This is one of nature's processes; and if all women were perfectly developed, and lived in harmony with nature's laws before and during preg-

nancy, there would be little need of artificial aids in parturition, and it would be attended with but little suffering. But, as it is, such aids are often needed; and, in cases where the natural powers become exhausted, nothing is so prompt and potent to restore natural and powerful uterine contractions as electro-magnetism, when properly and scientifically applied. This I know by practical experience. And, when we consider its power to arrest hemorrhage, we cannot help feeling surprised that our medical brethren should have paid so little attention to this subject, seeing that so many precious lives might be saved, and much suffering prevented, by a timely and proper use of this powerful agent.

Treatment.—To induce uterine contraction, apply the womb director with a piece of delicate moist sponge attached, negative S. C., to the os-uteri, with the positive on the abdomen over the fundus. Apply it for as long a time as the pains ought to continue, then resting until time for another pain. In this way natural uterine contraction will soon be established; and the labor will keep progressing to a successful termination, provided the presentation is natural, and there is no real want of capacity or deformity of parts. If profuse hemorrhage should follow, it must be treated as in cases of flooding.

Nothing aids so much in restoring tone to the whole system, as a little tonic treatment a few

days after delivery, and continued until perfect restoration to full tone and vigor.

DEFICIENT SECRETION OF MILK.

To promote the secretion of milk, apply the positive P. C. to the spine above the shoulders, and treat with the negative over the breasts from above downward; treat for ten minutes, two or three times a day.

SUPPRESSION OF MILK.

This is to be treated as above. Let the diet be generous, using plenty of milk in both cases. Warm, moist compresses applied frequently to the breasts aid the secretion.

SORE NIPPLES.

Apply the positive by means of the eye instrument, filled with diluted tincture of arnica, to the nipple, with negative P. C. between the shoulders. If the nipples crack, apply sweet cream. Bathe the nipples with diluted tincture of arnica, immediately after nursing, carefully washing it off with milk and water before nursing again. A solution of borax is good to harden the nipples, and aids in removing the soreness.

AGUE IN THE BREAST.

This painful condition of the breasts is generally caused by cold or some obstruction.

Treatment.—If there is general fever, with headache, place the negative at the feet, and treat all over with the positive, beginning at the base of the brain; and, for local treatment, treat over the breasts with the positive P. C., and negative on the spine a little below the positive.

GATHERED BREAST.

Treat as for ague in the breast until all the hardness and swelling subsides. Compresses of diluted arnica are serviceable in both conditions, both to take out the soreness and reduce inflammation.

ABSCESSES OF ALL KINDS.

Lumbar and other abscesses commence with inflammation, either spontaneously or from disturbing causes or local injuries.

Treatment.—Apply the positive to the swollen or inflamed part, S. C., and treat with the negative at some distant and convenient point. If suppuration cannot be prevented, it may hasten the cure to probe the swelling at the softest and most prominent point, and allow it to discharge. While this is going on, treat for a few times *over* and *around* the swelling; then treat as at first until well.

If it arises from scrofula, give the galvanic and bath treatment as soon as convenient, to eliminate the morbid elements from the system. If the

swelling causes general fever, alternate the local with general treatment, as in case of fever. In lumbar abscess, there is sometimes swelling in the groin, which discharges, and cures the abscess: this has sometimes been mistaken for hernia. In treating swellings and inflamed parts with the positive, it is well to have the negative on some part below the positive, because it is more soothing to the parts to run the current with the nervous ramifications than against them, or in the opposite direction,—just as it makes less commotion in the water when sailing with the current than when rowing up stream.

VARICOSE, OR ENLARGED VEINS.

Varicose veins during pregnancy, when in the limbs, are caused by the impregnated uterus pressing upon the vena-cava, thereby obstructing the free return of the blood to the heart. But other causes produce similar effects; yet all are to be treated on the same principles.

Treatment.—Apply the positive, S. C., a little below the varicose parts, with negative a little above, and pass them up the limb, a short distance from each other, toward the heart. The veins being relaxed, require the contracting influence of the positive. It is well sometimes, in addition, to apply bandages or laced stockings, and to occupy a recumbent posture, avoiding exercise as much as possible, and bathing the limbs frequently in cold

water. In chronic cases of men, it requires some time to make a permanent cure. If it is caused by the blood being too thick, that must be changed by proper food and drink, plenty of pure air, and frequent bathing. In addition to the above, the negative may be placed on the opposite side of the limb, while the positive is pressed freely on the enlarged veins, and both carried up as before.

NURSING SORE MOUTH.

When not constitutional, this depends on a disordered condition of the digestive organs and salivary glands.

Treatment.—Apply the positive to the tongue, P. C., and the negative on the stomach, liver, and spleen, for ten minutes. Then seat on the negative, and treat with the positive over the kidneys, liver, spleen, stomach, and intestines, from the duodenum to the lower part of the descending colon; and, if there is constipation, give tepid injections of soft water, until permanent relief is obtained, which will be as soon as the digestive organs regain their tone.

PUERPERAL OR CHILD-BED FEVER, AND INFLAMMATION OF WOMB AND OVARIES.

The conditions that mark this disease are inflammation of the womb and its appendages (fallopian tubes and ovaries), the peritoneum (the

lining membrane of the abdomen), or bowels, or all three, accompanied with general fever. This is a dangerous disease.

Treatment.—For local treatment, if convenient, place the patient in a full, warm sitz-bath, positive P. C. in the bath, and negative at the feet in warm water. But, if this is not convenient, treat over the bowels, ovaries, and womb with the positive sponge electrode, and negative at the feet, mild current, ten to twenty minutes; and, for general fever, treat with the positive from the head down to the feet, according to the necessity of the case, repeating the treatment every few hours until the symptoms improve. If there is constipation, relieve by tepid soothing injections, and observe all the rules of diet, as in other fevers. Be sure to prevent accumulation of milk in the breasts; keep the room well ventilated, and quiet as possible; and have the patient avoid all effort to move about. Keep the head cool, by moist cloths wrung out of cold water applied to the forehead and temples, and the disease will soon be subdued. Give occasional tonic treatment.

PHLEGMASIA DOLENS, OR MILK-LEG.

Treatment.—Place the negative at the feet, and treat with the positive over the affected parts, and the limbs generally, P. C., giving tonic treatment as soon as convenient, to restore the normal tone to all the organs and nervous system generally.

ULCERATION OF THE WOMB.

If from scrofula, treat generally as for that condition, and also apply the positive by means of womb instrument, P. C., to the os-uteri, and negative on the spine above the lumbar region.

ENLARGEMENT OF THE WOMB.

If from engorgement, apply the positve S. C. internally by womb instrument, and also over the womb, and negative on the spine, liver, spleen, and kidneys. Treat daily: keep the bowels regular, and see that the menses are natural and regular.

INDURATION, OR HARDENING OF THE WOMB.

If the organ is contracted, apply the negative P. C. both internally and externally, over the organ and ovaries, with positive on the spine, and over the kidneys, stomach, liver and spleen. Also sit the patient in a warm sitz-bath with the negative, and treat with positive as before.

PROLAPSUS UTERI, OR FALLING OF THE WOMB.

Antiversion, Retroversion, and Inversion, or Prolapsus of the Vagina.

This is one of the most annoying and formidable derangements to which woman is liable; and any means which are available, and reliable as a cure, must be a real blessing to every one suffering from

that cause. We are bold to affirm that electropathy, as a means to that end, is unparalleled, not only for falling of the womb, but for all derangements peculiar to that sex. Volumes might be written, descriptive of all the means used and theories advanced for the relief and cure of this and kindred disorders, and yet woman's sufferings continue; and this will continue until woman is better acquainted with her own functions, and obeys the laws which govern the healthy action of these functions.

Among the most prominent causes which induce morbid conditions are tight lacing, and heavy skirts wholly suspended from the waist, thus compressing the chest, and pressing downward all the viscera of the abdomen and pelvis. This produces relaxation and weakness of all the muscles and tissues, and prolapsus, antiversion, retroversion, and inversion, or what is called prolapsus of the vagina, which together make the worst form of prolapsus uteri. There are other causes which tend to produce these results, such as constipation, improper food, nervous excitement, amative indulgence, over-exertion, too long standing on the feet, injuries in childbirth, and want of proper exercise: young and old are both subject to these derangements. Whatever exhausts vitality in a woman may be a cause of prolapsus uteri. To live aright, dress aright, and refrain from all causes of exhaustion, observing every condition of health, are

amongst the means of a speedy and permanent cure. There is never prolapsus without dyspepsia and general debility, besides innumerable other symptoms too numerous to mention.

Treatment. — In curing prolapsus uteri, we restore the organ to its normal position; and, that being accomplished, we cure antiversion, retroversion, and inversion, or prolapsus of the vagina, for none of these can exist without prolapsus uteri, and relaxation of the walls of the vagina.

Now, as dyspepsia, general debility, and nervous prostration, generally accompany these affections, local treatment for dyspepsia, and general tonic treatment, must accompany the treatment for prolapsus. To treat prolapsus and its kindred ailments successfully, we must bring all the relaxed parts under the influence of the positive or contracting pole. We therefore apply the large silver or vaginal instrument internally, and press the womb up to its proper and natural position, connecting it with the positive pole, using the S. C., and at the same time apply the negative with the large sponge-handle on the spine, and over the stomach, just above the diaphragm. Have the current as strong as can well be endured. Treat for fifteen minutes in this way; then have the patient lie on her back, place the negative on the seventh cervical vertebræ, and withdraw the vaginal instrument, treating with the positive over the womb, ovaries, sides, and spine, as high as the

hips, and over the abdominal muscles generally. The patient ought to keep a recumbent position for several hours after each treatment, and avoid being much on her feet until the parts have become firm, and the organs remain permanently in their proper places. Cold vaginal injections, cool sitz-baths, and cold bathing of the abdomen, often aid in the cure, especially if there is irritation of any of the internal surfaces, or the womb, or ovaries, &c. Treat once a day, or alternate with the other treatment. This treatment persevered in will cure ninety-nine in a hundred, though of years standing.

BILIOUSNESS.

The above term is generally made use of to express diseases and conditions arising from a derangement of the liver and biliary organs, and is applied to colic, diarrhœa, fevers, &c., attended with general biliary derangement, and sometimes dyspepsia.

Make a thorough diagnosis, and treat generally, and locally according to the causes and symptoms. See Inflammation, Enlargement, Torpid Liver, &c., also Jaundice.

FISTULA IN ANO.

This is an orifice or opening from a cavity, abscess, or a local inflammation in the vicinity of the rectum; discharging pus, or thin, gleety, disordered matter, and not disposed to heal.

Treatment.—If there is an opening into the rectum, pass the rectum instrument, well oiled, into the rectum. Positive P. C. with negative all around the fistula outside. Let the current be gentle; treat once or twice daily, and, if it begins to heal, continue till well. Treat on the same principles as for ulcers. Keep the bowels free by injections, if necessary; avoid violent exercise, very salty food, and all stimulants.

SALT RHEUM.

This disease manifests itself by a dry eruption, forming fissures, caused by the skin becoming dry and cracking, from which a sharp, watery serum exudes, quite corrosive in its character.

Treatment.—The same as in scrofulous sores and skin affections. The pure galvanic current is the best, or the A B current on Dr. Jerome Kidder's machine, or Primary of Mr. Thomas Hall's machine.

SPERMATORRHŒA, OR SEMINAL WEAKNESS.

This disease results from excessive secretion and discharge of semen, and is caused by masturbation, excessive venery, or nocturnal emissions. Among the earlier and prominent symptoms of this disorder, there is a marked impairment of the mental faculties; the patient is unable to concentrate his mind upon his business or books; his memory becomes treacherous, his physical powers are weakened, his courage and energy fail, he is

languid and debilitated, becomes hypochondriacal and misanthropic, has fearful forebodings, is dyspeptic, loses flesh, sometimes becomes extremely emaciated; and all this without evidence of real organic disease. In this disease, there is irritation established in or about the seminal ducts, which convey the secretion from the testicle. There are also other causes of irritation, such as gonorrhœal lesions, mechanical obstructions, irritants in the rectum pressing upon the vesicular seminales or prostate gland, hemorrhoidal fissures, prostatitis, or even from stricture of the urethra.

It is not always easy to ascertain the true cause of the irritation on which the seminal flow depends. The patient often dislikes to make a frank disclosure of the facts concerning his own case, though the essential ones may be known to him alone; yet a correct knowledge of the true causes often aids in the successful treatment of the disease. The physician should strive to gain the unbounded confidence of his patient, and, having gained it, should regard it as a sacred trust, never to be violated.

Masturbation must be entirely abandoned, at any sacrifice, or it may end in insanity, dementia, or catalepsy. The cheerful and grateful influences of good society are indispensable; but, unfortunately for the sufferer, there is a disposition to avoid society of all kinds, especially that of females.

How lamentable the thought, that, for the lack of a little knowledge imparted to youth at the right time, tens of thousands are suffering all the terrible consequences of their ignorance, in impaired and broken-down constitutions; all their hopes of future usefulness and happiness entirely blasted; and, what is equally deplorable, not all the skill, aided by all the prescribed drug remedies of the schools, has yet proved adequate to cure the disease; and the poor sufferer, as he has failed under one, has fled to another, only to be disappointed, and bereft of all hope of ever finding relief.

Happily for the sufferer, this does not apply to our mode of treating this and all kindred diseases, as thousands can testify, whom we have entirely cured after all other means had been tried in vain. This we do not say boastingly; but we feel bold to affirm what we know to be true, and feel prepared to prove by actual demonstration.

Treatment.—As there is frequently emaciation and general debility accompanying this disease, it is well to give general tonic treatment for a few days or a week at first, using S. C., and commencing with the current very light, finishing this part of the treatment by a few passes on the cerebellum or back brain, manipulating with both hands but *without* the current, over all the head.

To remove irritation from the genital organs, and proximate parts, place the positive under a

large sponge in a shallow sitz-bath, of a temperature to be comfortable to the patient. Treat with the negative up the spine to between the shoulders, and over the stomach, liver, and spleen: have the current pleasant to the patient; treat in this way for ten to fifteen minutes, rub dry, and then finish with a general rubbing with the dry, warm hand. If the irritation is confined to the urethra and testicles, the testicles and penis may be inserted with the positive in a cup with tepid water, and the negative applied as before; or if the irritation is confined to the rectum, the patient may be seated on the positive, or the rectum instrument may be used. After the irritation has been subdued, and we desire to give tone to all the parts involved, treat with the positive on the spine from the back brain downwards, and over the kidneys, negative in the bath, or cup, or seated on the sponge as before. It requires from one week to three months to cure these cases; but an improvement will be experienced almost from the beginning in the worst cases, and often in recent cases a cure can be effected in a week. Cold compresses, cold sitz-baths, and frequent bathing of the parts in cool or cold water, both allays the irritation, and aids in giving tone to all the organs and parts involved. The diet may be nutritious, but not stimulating. All stimulants must be avoided, and the exercise taken must be gentle and

at regular intervals. If there is dyspepsia, or any other *distinct* disease, treat accordingly.

IMPOTENCE.

This condition is generally induced by excessive sexual intercourse. Married men, both young and middle-aged, with others, often find themselves in this unpleasant dilemma. I have cured many of this class without one particle of medicine.

Treatment.—First cease to attempt sexual intercourse until the power is restored. If the general health and tone of the system is impaired, commence by a course of general tonic treatment. Afterward seat the patient on the negative, either on the sponge or in a cool sitz-bath, or place the genitals in a cup, S. C., and treat with the positive on the spine, back brain, and over the kidneys and abdomen, with a moderate current. In treating on the spine, treat freely over the lumbar region; treat daily for ten to fifteen minutes. Let the diet be nutritious, but not stimulating. Bathing as in spermatorrhœa, with moderate exercise in the open air.

SWELLING AND INFLAMMATION OF THE GENITALS.

Treatment.—Place the organs in a cup or sitz-bath, with the positive S. C., and treat with the negative up the spine, over and above the lumbar region. Treat several times daily, if necessary, or according to the urgency of the case. Frequent

cool or cold sitz-baths are good; let the diet be cooling, and avoid walking; keep a recumbent position as much as possible. This treatment may be varied by placing the negative at the feet.

SWEATING OF THE GENITALS.

Treatment.—Place the organs in a cup or sitz-bath, with negative, and treat with positive P. C. over the kidneys, spine, and abdomen. Give general tonic treatment if needed, and have the patient take a general bath three times a week, and a cool sitz-bath on alternate days. Keep the bowels regular, and avoid all extremes of eating, drinking, and exercise: treat daily from ten to fifteen minutes.

SYPHILIS, OR VENEREAL DISEASE.

This loathsome disease shows itself in every variety of form; from mere primary symptoms, to the most foul and disgusting ulcers, affecting not only the genitals and other soft parts, the eyes, nose, mouth, throat, and skin, but also producing caries of the bones. It is caused by a poisonous, infectious principle, which can only act from contact, the smallest particle of which when brought in contact with an abraded mucous surface, is sufficient to produce a local disease, and, from its absorption, to affect the entire system. Locally, the disease consists of chancres, or venereal ulcers, which suppurate; and, if the infectious matter is

re-absorbed, the blood becomes poisonous, and the whole train of secondary syphilitic affections are liable to follow.

Treatment.—In treating this disease, the pure galvanic current is preferable to any other; but the primary current of the machine answers very well. In treating with the galvanic, six to eight cups may be used. Treat on the part affected with the positive; and, if on the penis, treat on the opposite side with the negative. As this current is chemical in its action, we depend on that to neutralize the virus of the ulcer; but, if it comes to suppuration, either cauterize with nitrate of silver, or with a paste made of sulphuric acid, mixed with powdered vegetable charcoal, in the proportions necessary to form a half-solid paste. It is proposed, by this application, to destroy the poison and convert the chancre into a simple wound, which will proceed rapidly to cicatrization. At the same time give homeopathic mercurius, third trituration, in some form, one grain night and morning, until the skin of the sore looks natural and healthy, and like the other skin. Treat with the galvanic current once or twice daily, as the case seems to require. If there is a swelling in the groin, treat over the swelling with the positive, and seat the patient on the negative; but, if suppuration takes place, use slippery elm and flaxseed poultice, giving the internal remedy, the same as before. Avoid all exposure to cold,

and let the exercise be gentle. Avoid all stimulants, and let the diet be soothing and nutritious, but free from stimulating spices.

SECONDARY SYPHILIS.

Secondary syphilis consists in the introduction of a poison into the blood; and the cure of it in the neutralization of that poison, and healing of the parts affected.

Treatment.—If it appears upon the skin, in sores or blotches of any kind, give the patient a general galvanic bath. Place the positive either at the feet, or at the base of the spine, and treat all over the surface with the negative, especially those parts occupied by the discolorations or sores. Treat from fifteen to thirty minutes; then rub dry, and, if the air is cool, keep the patient in-doors for half an hour. Give this treatment every other day, and on alternate days treat locally, either with the primary current or the pure galvanic.

If the throat is affected, treat with the positive in the mouth, and negative on the spine. Treat all external sores with the positive or negative, according to their positive or negative condition. Remember that it is more important to effect a chemical change in treating this disease, than to induce polar action. Give homeopathic mercurius, one grain every other day, and sulphur on the alternate days, until well. The great majority of

these cases are curable in a short time : we seldom fail in any.

NODES.

It is all important that these hardened tumors should be removed as soon as possible, and suppuration prevented.

Treatment.—Apply the positive P. C., or the pure galvanic to the node, and either seat on the negative, or place it at the feet. Vary the treatment by treating through and through, giving the internal remedies, as in secondary syphilis.

GONORRHŒA.

This is a purulent, greenish-yellow discharge from the urethra in males, and from the vagina in females, attended with heat, swelling, and inflammation of the parts, with burning and scalding during micturition, and painful erections.

Treatment.—Seat the patient on the positive P. C., in a tepid shallow sitz-bath, and apply the negative on the spine, over the dorsal vertebra, for fifteen to thirty minutes; or place the metallic band around the body, over the liver, and kidneys, with negative and positive in the bath; or place the genitals in a cup, with positive and negative as above. Use the vaginal instrument, same pole, for females. Give homeopathic mercurius cor. daily, in one grain doses, third trituration. Should the discharge continue after most of the local inflam-

mation has subsided, make use of a weak solution of nitrate of silver, as an injection; and for females, a solution of soluble mercury in water, as a vaginal injection. Inject two or three times daily.

Avoid all high-seasoned food, and stimulating beverages. Mucilaginous drinks of slippery elm and gum-arabic are best. A recumbent posture is best: take no more exercise than is actually necessary. Preserve perfect cleanliness, and be careful not to communicate, by means of the fingers or towels, any of the virus to the nose, eyes, or mouth, or any abraded surface, as it will similarly affect those parts, producing purulent discharges, &c.

GLEET.

This follows frequent attacks of gonorrhœa, and the debility of the parts occasioned thereby. This disease is uninfectious, and unattended with pain, unless there is stricture.

Treatment.—Give, first, general tonic treatment; next place the genitals in the cup, with the positive S. C., negative on the spine, and over the kidneys. Finish with negative in the cup, positive on spine and cerebellum. Treat daily, ten to fifteen minutes. Give occasional cold sitz-baths, and bathe the genitals daily in cold water. Diet liberal, but not stimulating, exercise moderate; avoid lifting and straining, and everything that tends to fatigue.

INDEX.

Abdomen, dropsy of, 121.
Abortion, 176.
Abscesses, 180.
Ague in the breast, 179.
Alopecia, 149.
Amaurosis, 92.
Amenorrhœa, 172.
Anasarca, 120.
Aphonia, 112.
Apoplexy, 130.
Apparatus, electrical, 60 ; Dr. J. Kidder's 68 ; Mr. Thomas Hall's, 64 ; magneto-electric, 65.
Arthritis, 103.
Ascites, 121.
Asiatic cholera, 108.
Asphyxia, 159.
Asthma, 139.

Back, crick in, 162.
Bath, galvanic, 69.
Battery, galvanic, 60, 64.
" Chester's electropion, 63, 67.
Biliousness, 187.
Biliary calculi, 86.
Bladder, inflammation of, 79 ; stone or gravel in, 168.
Blood, circulation of, 162 ;
" voided by urine, 140.
Boils, 160.
Brain, dropsy of, 122 ; inflammation of, 80.
Breast, ague in, 179 ; gathered, 180.

Bright's disease of kidneys, 170.
Bronchia, inflammation of, 98.
Bronchitis, 98 ; chronic, 99.
Bruises, 160.
Bunions, 161.
Burns, 114.

Calculi, biliary, 86; urinary, 168.
Cancer, 124.
Canker in mouth, 161.
Catalepsy, 136.
Catarrh, 114 ; chronic, 115.
Cataract, 91.
Cephalagia, 151.
Cessation of menses, 175.
Change of life, 175.
Chest, dropsy of, 121.
Chester's electropion battery, 62, 68.
Chicken Pox, 105.
Chilblains, 114.
Chlorosis, 172.
Cholera Morbus, 107.
" Asiatic, 108.
" Infantum, 109.
Chorea, 133.
Colica, or colic, 112.
Common colds, 115.
Conditions of health and disease, 33-41.
Congestion, 106.
Constipation, or Costiveness, 112.
Consumption of lungs, 117.
Contusions, 160.
Corns, 161.

Crick in neck, back, limbs and stomach, 162.
Croup, 97.
Curative agents, 41-46.
Cystitis, 79.

Deafness, 150.
Debility, general, 163; nervous, 163.
Deficient secretion of milk, 79.
Delirium tremens, 147.
Delivery, 177.
Derangements, mental, 145.
Diabetis, 157.
Diagnosis, electrical, and medication, 71-78.
Diaphragmitis, 89.
Diaphragm, inflammation of, 89.
Diarrhœa, acute, 110; chronic, 110.
Diphtheria, 97.
Disease, conditions of health and, 33-41.
Disease, philosophy of treating, 47-59.
Dropsy of limbs and feet, 120; of abdomen, 121; of chest, 121; of brain, 122; ovarian, 122; of testicles, 123.
Dysentery, 112.
Dysmenorrhœa, 173.
Dyspepsia, 110.

Ears, ringing in, 152, frozen 162.
Electrical apparatus, 60.
" machine, 60.
" currents, 65.
" diagnosis and medication, 71-77.
Electropian battery, Chester's
" 63, 70.
" to make, 63.
Electro-magnetic or Faradaic machine, 64, 67.
Electrodes, for local treatment, 68.

Enlargement of the womb, 184.
Enteritis, 87.
Epilepsy, 129.
Eruptive fever, 105.
Eruptions, 105.
Erysipelatous inflammation, 78.
Erysipelas, 144.
Eye, inflammation of, 90; lids thickened or granulated, 91.

Fainting, 137.
Falling of the womb, 184.
Falling off of the hair, 149.
Falling sickness, 129.
Far-sightedness, 94.
Felons, 160.
Fever and ague, 104.
Fever, remittent, 105, intermittent, 105; eruptive, 105; scarlet, 105; puerperal, or childbed, 182.
Fistula lachrymalis, 93; in ano, 187.
Flooding, 174.
Fluor albus, 173.
Frost bitten limbs, 162.
Frozen ears, limbs, &c., 162.

Galvanic battery, 61, 66.
" bath, 69.
Gastritis, 84.
Gathered breast, 180.
General tonic treatment, 75.
Genitals, inflammation of, 192 sweating of, 192.
Giddiness, 137.
Glandular enlargement, 126.
Gleet, 197.
Glossitis, 97.
Gonitis, 103.
Gonorrhœa, 196.
Gout, 103.
Gravel, 168.
Green sickness, 172.
Gums, inflammation of, 100.

Hall, Mr. Thomas, electro medical apparatus, 65.

Hardening of womb, 184.
Harmonious growth of the physical, mental, and spiritual, necessary for the health of man, 23-27.
Head, noises in, 152.
Headache, nervous, 151 ; sick, 152.
Health and disease, conditions of, 33-40.
Heamaturia, 140.
Heart, palpitation of, 137 ; enlargement of, 137 ; weakness of, 137.
Hemiplegia, 132.
Hemorrhoids, 141.
Hemorrhagia, active, 125 ; passive, 124.
Hepatitis, 85.
Hernia, 143.
Hip disease, 104.
Hydrops, 119.
Hydrothorax, 121.
Hydrocephalus, 122.
Hydrocele, 123.
Hypertrophy, 137.
Hypochondria, 147.
Hysteria, or hysterics, 148.

Influence of mind over man, 27-32.
Instruments, or electrodes, for local treatment, 68.
Inflammation, 78 ; phlegmonous, acute, chronic, erysipelatous, 78 ; of bladder, 79 ; of brain, 80 ; of lungs, 82 ; of pleura, 83; of stomach, 84 ; of liver, 85 ; of intestines, 87 of kidneys, 88 ; of spleen, 88; of diaphragm, 89 ; of the eye, 90 ; of the peritoneum, 94 ; of the internal ear, 94 ; of larynx, 95 ; of tonsils, 96 ; of tongue, 97; of bronchia, 98 ; of palate, 100 ; of gums, 100 ; of knee, 103 ; of the ovaries, 122 ; of the womb, 182 ; of genitals, 192.
Intermittent fever, 105.
Icterus, 113.
Influenza, 115.
Intestines, inflammation of, 87.
Immoderate flow of the menses, 174.
Irregular menstruation, 175.
Induration, or hardening of womb, 184.
Impotence, 192.

Jaundice. 113.

Kidder, Dr. Jerome, apparatus, 68.
Kidneys, inflammation of, 88 ; Bright's disease of, 170.
Knee, inflammation of, 103.

Local treatment, electrodes for, 63.
Lungs, inflammation of, 82 ; hepatization of, 156; consumption of, 156 ; weakness of, 166.
Liver, inflammation of, 85 ; enlarged, 86 ; torpid and hardened, 86.
Larynx, inflammation of, 95.
Laryngitis, 95.
Limbs, frost-bitten, 162 ; dropsy of, 120 ; frozen, 162, cramp in, 162.
Loss of voice, 116.
Life, change of, 175.
Leucorrhœa, 173.

Matter, primal condition of, 9-14.
Man, continued progress of, 15-22.
" health of, dependent upon harmonious growth, 23-26.
" influence of mind over 27-32.

202 INDEX.

Mind, its influence over man, 27-32.
Medication, 75.
Magneto-electric apparatus, 65.
Myopia, 93.
Measles, 105.
Mental derangements, 145.
Mania-a-potu, 147.
Mouth, sore, 161; nursing sore, 182; canker in, 161.
Menses, retention of, 171; suppression of, 172; immoderate flow of, 174; cessation of, 175.
Menstruation, irregular, 175; painful or difficult, 173.
Menorrhagia, 174.
Miscarriage, 176.
Mumps, 101.
Milk, deficient secretion of, 179; suppression of, 179.
Milk leg, 183.

Nephritis, 88.
Near-sightedness, 93.
Neuralgia, 156.
Nervous headache, 151.
Nervous debility, 163.
Noises in head, 152.
Nipples, sore, 179.
Nursing sore mouth, 182.
Nodes, 196.
New-born infants, asphyxia of, 160.

Ovaries, inflammation of, 122.
Opthalmia, 90.
Otitis, 94.
Ovarian dropsy, 122.
" tumor, 122.
" inflammation, 122.
Oreana, 164.

Primal condition of matter, 9-14.
Progress of creation up to man, 9-14.
" of man as a physical and spiritual being, 15-22.
Philosophy of treating disease, 47-59.
Phlegmonous inflammation, 78.
Pleura, inflammation of, 83.
Peritoneum, inflammation of 94.
Palate, inflammation of, 100.
Phrenitis, 80.
Pneumonia, 82.
Presbyopia, 94.
Peritonitis, 94.
Paratitis, 101.
Puerperal fever, 182.
Phthisis pulmonalis, 117.
Paralysis, 131; local, 132.
Palsy, 131.
Palpitatio cordis, 137.
Palpitation of the heart, 137.
Piles, 141.
Prolapsus ani, 142; uteri, 184.
Polypus, 164.
Pregnancy, 176.
Parturition, 177.
Phlegmasia dolens, 183.

Quinsy, 96.

Remarks upon electrical currents, 65.
Rheumatism, inflammatory, 101; chronic, 102.
Remittent fever, 105.
Retention of urine, 140; of menses, 171.
Rupture, 143.
Rickets, 144.
Ringing in ears, 152.
Recent wounds, 160.

Spiritual, mental and physical growth, 23-32.
Stomach, inflammation of, 84; cramp in, 162; weakness of, 166.
Spleen, inflammation of, 88.
Splenitis, 88.
Stye, 91.

Strabismus, 93.
Squinting, 93.
Scarlet-fever, 105.
Small-pox, 105.
Sores and ulcers, 105.
Scalds, 114.
Spermatocele, 123.
Swellings, 126.
Scrofula, 127.
St. Vitus' dance, 133.
Syncope, 137.
St. Anthony's fire, 144.
Spinal curvature, 153; irritation, 155.
Suspended animation, 159.
Sore mouth, 161; nursing, 182.
Sore nipples, 179.
Sick headache, 152.
Sleeplessness, 165.
Stammering, 165.
Sunstroke, 166.
Stone in the bladder, 168.
Suppression of menses, 172; of milk, 179.
Salt rheum, 188.
Spermatorrhœa, 188.
Swelling of the genitals, 192.
Sweating of the genitals, 193.
Syphilis, 193; secondary, 195.

Treating disease, philosophy of, 47–59.

Treatment, 78–197.
" general tonic, 75.
Tonsils, inflammation of, 96.
Tongue, inflammation of, 96.
Tonsilitis, 96.
Trachitis, 97.
Testicles, dropsy of, 123.
Tumors, 122.
Tetanus, 135.
Toothache, 167.

Urine, retention of, 140; voiding blood by, 140.
Ulcer of the nose, 164.
Urinary calculi, 168.
Utero-gestation, 176.
Ulceration of the womb, 184.

Vertigo, 137.
Voiding blood by urine, 140.
Vomiting, 166.
Varicose veins, 181.

Womb, inflammation of, 182.
  ulceration of, 184; enlargement of, 184; hardening of, 184; falling of, 184.
Wounds, recent, 160.
Weakness of stomach, lungs, heart, 166.
Water-brash, 166.
Whites, 173.

Phrenology and Physiognomy.

ological Jour-
TRATED. Publish-
year; 30 cts. a No.
renology and
yearly 12mo vol.
the current year.
8, '69, '70, '71, '72,
; over 350 Illustra-
e. $2.00.
Ian.—Considered
ernal Objects. By
uslin, $1.50.
of Phrenolo-
s, $1.50.
g various Develop-
for Phrenologists.
iology; Contain-
Nature and Value
Evidence. By Dr.
, $1.25.
r, Marriage Vindi-
e Exposed. By N.
elementary Princi-
e Nature of Man.
, M.D. $1.25.
Self-Improve-
Comprising Physi-
Mental; Self-culture
tellectual Improve-
.50.
Anatomy and Phi-
HARLES BELL, K.H.
tes, and upward of
s. $1.25.
aracter.—A New
ook of Phrenology
for Students and
Engravings. $1.00;
hrenology.—By
Notes. $1.50.
story and Ceremo-
logical and Physio-
f the Functions and
Happy Marriages.
Phrenology and
to the Selection of
ons for Life. 50 cts.
ntellectual Im-
l to Self-Education
ection. $1.25.
, according to the
enology. By G. S.

Moral Philosophy; or, the Duties of Man considered in his Individual, Domestic and Social Capacities. By GEORGE COMBE. $1.50.

Natural Laws of Man.—Questions with Answers. By J. G. SPURZHEIM. Muslin, 50 cents.

New Physiognomy; or, Signs of Character, as manifested through Temperament and External Forms, and especially in the "Human Face Divine." With more than One Thousand Illustrations. By S. R. WELLS. In three styles of binding. Price, in one 12mo volume, 768 pp. Muslin, $5; heavy calf, marbled edges, $8; Turkey morocco, full gilt, $10.

Phrenology and the Scriptures. By Rev. JOHN PIERPONT. 25 cents.

Phrenology Proved, Illustrated, and Applied. $1.50.

Phrenological Busts.—Showing the latest classification, and exact location of the Organs of the Brain, fully developed, designed for Learners. It is divided so as to show each individual Organ on one side; and all the groups—Social, Executive, Intellectual, and Moral—properly classified on the other side. There are two sizes. The largest is sold in Box, at $2; The smaller one, by mail, post-paid, $1.

Phrenological Guide. Designed for the Use of Students. Paper, 25 cts.

Self-Culture and Perfection of Character, with Management of Children. $1.25.

Self-Instructor in Phrenology and Physiology. With over One Hundred Engravings. Paper, 50 cents; muslin, 75 cents.

Symbolical Head and Phrenological Map, on fine tinted paper. 10c.

Wells' New Descriptive Chart for Use of Examiners, giving a Delineation of Character. Paper, 25 cents; Flexible muslin, 50 cents.

Your Character from Your Likeness. For particulars, how to have pictures taken for examination, inclose stamp for a copy of "Mirror of the Mind."

to the "SCIENCE OF MAN," including Phrenology, Physiognomy,
y, Physiology, Anatomy, Hygiene, Dietetics, etc., supplied.
ated and Descriptive Catalogue with full titles, also inclose stamp
gents.

WELLS & CO., Publishers, 737 Broadway, N. Y.

New and Standard Health Books.

Anatomical and Physiological Plates. Arranged expressly for Lectures on Health, Physiology, etc. By R. T. TRALL, M.D. Six, fully colored, and mounted on rollers. Price, $15, net.

Accidents and Emergencies.—A Guide for Treatment of Wounds, Burns, Sprains, Bites, Drowning, etc. New and revised edition. 25 cents.

Alcoholic Controversy.—A Review of the *Westminster Review*. 50 c.

Chemistry, and Its Application to Physiology, Agriculture, and Commerce. By LEIBIG. 25 cents.

Children.—Their Management in Health and Disease. By SHEW. $1.50.

Cure of Consumption by the SWEDISH MOVEMENTS. By Dr. WARK. Paper, 25 cents.

Digestion and Dyspepsia.—The Digestive Process explained, and Treatment of Dyspepsia given. $1.00.

Diseases of the Throat and LUNGS. Illustrated. 25 cents.

Domestic Practice of Hydropathy. By E. JOHNSON, M.D. $1.50.

Family Gymnasium.—Methods of applying Gymnastic, Calisthenic, and Kinesipathic Exercises. $1.50.

Food and Diet.—With Observations on the Dietetical Regimen, suited for Disordered States of the Digestive Organs. By PEREIRA. $1.50.

Fruits and Farinacea, the PROPER FOOD OF MAN. $1.50.

Health Catechism or Questions AND ANSWERS. By Dr. TRALL. 10 cts.

Hydropathic Cook-Book.—A Philosophical Exposition of the Relations of Food to Health. $1.25.

Hydropathic Encyclopedia.—Embracing Anatomy, Physiology of the Human Body; and the Preservation of Health, including the Nature, Causes, Symptoms, and Treatment of Disease. By R. T. TRALL, M.D. $4.00.

Hygeian Home Cook-Book; or, How to Cook Healthful and Palatable Food without Condiments. 25 cents.

Family Physician.—A Ready Prescriber and Hygienic Adviser. $4.00.

Management of Infancy, Physiological and Moral Treatment. $1.25.

Medical Electricity, showing its most scientific application to all forms of Disease. By WHITE. $2.

Midwifery and the Diseases of WOMEN. With General Management of Child-birth, the Nursery, etc. $1.50.

Movement-Cure.—An Ex[] of the Swedish Movement-Cu[]

Mother's Hygienic H[] BOOK, for the Normal Developm[] Training of Women and Childr[] the Treatment of their Disease[]

Notes on Beauty, Vigor DEVELOPMENT; or, How to Plumpness, Strength, and Beau[]

Philosophy of the Water[] A Development of the Princ[] Health and Longevity. 50 cent[]

Popular Physiology.—A [] the general reader, and arrang[] Text-book for Schools, Colleg[] Families. $1.25.

Physiology of Digestion relation to the principles of D[] By ANDREW COMBE, M.D. 50 c[]

Practice of the Water-C[] An account of the various proce[]

Physiology, Animal and TAL. Health of Body and of Min[]

Principles of Physiology to the Preservation of Health Improvement of Physical and Education. By COMBE. $1.50.

The Science of Human L[] SYLVESTER GRAHAM. $3.00.

Sober and Temperate Discourses and Letters of Lo[] NARO, with Biography. 50 cen[]

Tea and Coffee, Their Phy[] tellectual, and Moral Effects. []

The Bath.—Its History and Health and Disease. Paper, 25[]

The Human Feet.—Their Dress, and Care. Muslin, $1.00[]

The True Healing Art; [] enic vs. Drug Medication. 2[]

The Parents' Guide; or, Development through Inherit[] dencies. $1.25.

The Hygienic Hand-Bo[] Guide for the Sick-Room. $1.[]

Tobacco. Its Physical, Inte[] and Moral Effects. 25 cents.

Water-Cure in Chroni[] EASES.—Causes, Progress and T[] tion of the Diseases, and Tre[] By J. M. GULLY, M.D. $1.50.

Water-Cure for the Mil[] Process of Water-Cure Explain[]

Special List.—We have also Medical Works and Treatises although not adapted to gener[] lation, are invaluable to th[] need them. This Special List[] sent on *receipt of stamp.*

For sale by Booksellers, or sent by mail, post-paid, on receipt of price. Agent[]

Address **S. R. WELLS & CO., Publishers, 737 Broadway**

Works for Home Improvement.

The Indispensable Hand-Book for Home Improvement. Comprising "How to Write," "How to Talk," "How to Behave," and "How to Do Business." One large vol., $2.25.

How to Write, a Manual of Composition and Letter-Writing. Muslin, 75c.

How to Talk, a Manual of Conversation and Debate, with Mistakes in Speaking Corrected. 75 cents.

How to Behave, a Manual of Etiquette and Guide to Personal Habits, with Rules for Debating. 75 cents.

How to Do Business, a Pocket Manual of Practical Affairs, and a Guide to Success, with Legal Forms. 75 cts.

Right Word in the Right Place. Dictionary of Synonyms, Technical Terms, Phrases, etc. 75 cents.

Weaver's Works.—Comprising "Hopes and Helps," "Aims and Aids," "Ways of Life." One vol., $2.50.

Hopes and Helps for the Young Character, Avocation, Health, Amusement, Courtship and Marriage. $1.25.

Aims and Aids for Girls and Young Women, on Duties of Life. $1.25.

Ways of Life, showing the Right Way and the Wrong Way; the Way of Honor and the Way of Dishonor. $1.

Life at Home; or, the Family and its Members. Husbands, Wives, Parents, Children, Brothers, Sisters, Employers and Employed. $1.50; gilt, $2.

Wedlock; or, the Right Relations of the Sexes, disclosing the Laws of Conjugal Selection. Showing Who May and Who May Not Marry. By S. R. WELLS. $1.50; full gilt, $2.

Oratory—Sacred and Secular; or, The Extemporaneous Speaker. Including a Chairman's Guide. $1.25.

The Temperance Reformation. From the first Temperance Society in the U. S. to the Maine Law. $1.25.

How to Paint.—Designed for Tradesmen, Mechanics, and Farmers. Plain and Fancy Painting, Graining, Varnishing, Kalsomining, and Paper Hanging. By GARDNER. $1.00.

The Carriage Painter's Illustrated Manual. A Treatise on the Art, Science, and Mystery of Coach, Carriage, and Car Painting. $1.00.

Man in Genesis and in Geology; or, the Biblical account of Man's Creation tested by Scientific Theories. $1.

Heart Echoes, a Book of Poems. By HELEN A. MANVILLE. Cloth, $1.

The Conversion of St. Paul.— By GEO. JARVIS GEER, D.D. 75 cents.

The Emphatic Diaglott; or, The New Testament in Greek and English. Containing the Original Greek Text of the New Testament, with an Interlineary English Translation. By BENJ. WILSON. Price $4; extra fine, $5.

The Culture of the Human Voice.—Its Anatomy, Physiology, Pathology, Therapeutics, and Training. By TRALL. 50 cts.; cloth, 75 cts.

Æsop's Fables Illustrated.— People's Pictorial Ed. Tinted paper, $1.

Gems of Goldsmith.—The Traveler, The Deserted Village, and the Hermit. With Illustrations. Full gilt. $1.

Pope's Essay on Man.—With Notes. Beautifully illustrated. Gilt, $1.

Library of Mesmerism and Psychology. Comprising the Philosophy of Mesmerism—Fascination.—The Macrocosm.—Electrical Pyschology.— The Science of the Soul. One vol. $3.50.

Fascination; or, the Power of Charming. By J. B NEWMAN. $1.25.

Salem Witchcraft, with Planchette Mystery and Modern Spiritualism, and Dr. Doddridge's Dream. $1.

Fruit Culture for the Million. A Guide to the Cultivation and Management of Fruit Trees. New Ed. 75c.

Saving and Wasting, or Economy Illustrated in a Tale of Real Life. $1.25.

Footprints of Life; or, Faith and Nature Reconciled. A Poem. $1.25.

A Self-Made Woman; or, Mary Idyl's Trials and Triumphs. $1.50.

Home for All, or the Gravel Wall. Showing the Superiority of Concrete over Brick, Stone, or Frame Houses, with Octagon Plans. $1.25.

Philosophy of Electrical Psychology. In Twelve Lectures. $1.25.

Philosophy of Mesmerism and Clairvoyance. Six Lectures, with Instructions. 50 cents.

Thoughts for the Young Men and Young Women of America. 75c.

The Christian Household.—Embracing the Husband, Wife, Father, Mother, Brother, and Sister. 75 cents.

Capital Punishment, or the Proper Treatment of Criminals. 10 cts. Education of the Heart. 10 cts. Father Matthew, the Temperance Apostle. 10c. Good Man's Legacy. 10 cts. Gospel among Animals. 10 cts. The Planchette Mystery—how to work it. 20c. Alphabet for Deaf and Dumb. 10c.

Temperance in Congress.—25 c.

Address **S. R. WELLS & CO.,** Publishers, 737 Broadway, N. Y.

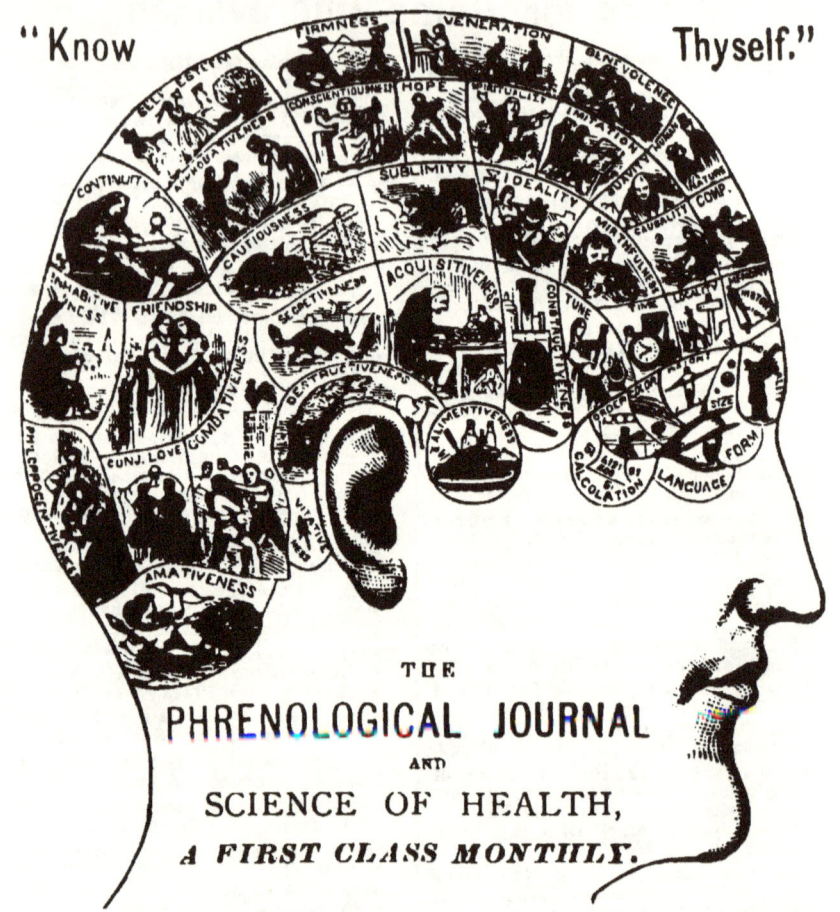

"Know Thyself."

THE PHRENOLOGICAL JOURNAL
AND SCIENCE OF HEALTH,
A FIRST CLASS MONTHLY.

Specially Devoted to the "SCIENCE OF MAN." Contains PHRENOLOGY and PHYSIOGNOMY, with all the "SIGNS of CHARACTER, and how to read them;" ETHNOLOGY, or the Natural History of Man in all his relations to Life; Practical Articles on PHYSIOLOGY, DIET, EXERCISE and the LAWS of LIFE and HEALTH. Portraits, Sketches and Biographies of the leading Men and Women of the World, are important features. Much general and useful information on the leading topics of the day is given. It is intended to be the most interesting and instructive PICTORIAL FAMILY MAGAZINE Published. Subscriptions may commence now.

Few works will better repay perusal in the family than this rich storehouse of instruction and entertainment.—*N. Y. Tribune.* It grows in Variety and Value. *Eve Post*

Terms.—A New Volume, the 63d, commences with the July Number. Published Monthly, in octavo form, at $3 a year, in advance. Sample numbers sent by first post, 20 cents. Clubs of ten or more, $2 each per copy, and an extra copy to agent.

We are now offering the most liberal premiums ever given for clubs, for 1876. Inclose stamps for list. Address. S. R. WELLS & CO., 737 Broadway, New York.

www.ingramcontent.com/pod-product-compliance
Lightning Source LLC
Chambersburg PA
CBHW020903230426
43666CB00008B/1298